Artistic Dialogues with the Arctic North

This book brings together Indigenous and non-Indigenous artists and scholars to explore how environmental changes in the Arctic are being experienced, understood, and represented. As climate change unfolds across vast geographies and long timescales, its impacts—on identity, wellbeing, cultural resources, and ways of life—are intimate, abrupt, and deeply felt.

Across the thirteen chapters, authors from diverse regions, disciplines, and artistic practices document and respond to these shifts, offering nuanced insights into the lived realities of a warming North. From ephemeral transformations in the land to the erosion of shared memory and subsistence traditions, these creative and scholarly responses serve as powerful forms of witnessing and response. Rather than relying solely on metrics or models, this volume emphasizes sensory, emotional, and cultural knowledge, foregrounding how individuals and communities are making sense of, and responding to, ecological disruption. It offers a vital perspective on the disproportionate and uneven impacts of climate change, and the urgent need to attend to its effects at both personal and planetary scales.

This is an essential resource for scholars, students, and practitioners in environmental history, Indigenous studies, climate policy, art and visual culture, and global development, and for anyone seeking new ways to engage with environmental change through art, experience, and collaboration.

Antonia Sohns earned her PhD in Geography from McGill University (2020). Her research and work focus on water security and climate change adaptation. She holds an MSc in Water Science, Policy and Management from the University of Oxford (2011) and a BSc in Earth Systems, Oceans track from Stanford University (2010). She has conducted environmental and social science research from the Arctic to the tropics, including for the Smithsonian Tropical Research Institute in Panama and for international NGOs.

Routledge Environmental History

The new *Routledge Environmental History* series presents current interdisciplinary research investigating our changing ecological entanglements over time. It draws on diverse time periods, geographical regions, methods and ideas from subject areas such as history, geography, archaeology, and the natural sciences. It taps into how topical debates such as the roles of non-humans and more-than-humans, social and environmental justice, climate change, and antiracism are shaping environmental histories and shifting approaches to engaging with the natural world.

Comprising edited collections, co-authored volumes and single author monographs, this collection provides an invaluable resource for advanced undergraduate and postgraduate students and scholars with an interest in the wide-ranging field of environmental history.

For more information, or to submit a proposal, please contact the Commissioning Editor, Grace Harrison (grace.harrison@tandf.co.uk).

A Cultural History of Waste Disposal
Environmental Policy and Park Redevelopments
Benjamin A. Lawson

The History of Environmental Degradation in Mar Menor
A Case Study
Edited by Juan Manuel Zaragoza, David Soto, and Malena Cantero

The History of Environmental Resource Management in Europe
Sustainable Practices Through Time
Edited by Matteo Di Tullio, Anna Maria Stagno and Martino Lorenzo Fagnani

Artistic Dialogues with the Arctic North
Environmental Change and Identity in Transition
Edited by Antonia Sohns

For more information about this series, please visit: www.routledge.com/Routledge-Environmental-History/book-series/ESREH

Artistic Dialogues with the Arctic North

Environmental Change and Identity in Transition

Edited by Antonia Sohns

Routledge
Taylor & Francis Group
LONDON AND NEW YORK

earthscan
from Routledge

First published 2026
by Routledge
4 Park Square, Milton Park, Abingdon, Oxon OX14 4RN

and by Routledge
605 Third Avenue, New York, NY 10158

Routledge is an imprint of the Taylor & Francis Group, an informa business

British Library Cataloguing-in-Publication Data
A catalogue record for this book is available from the British Library

ISBN: 978-1-032-90030-8 (hbk)
ISBN: 978-1-032-90032-2 (pbk)
ISBN: 978-1-003-54588-0 (ebk)

DOI: 10.4324/9781003545880

Typeset in Times New Roman
by SPi Technologies India Pvt Ltd (Straive)

Contents

Contributors

Eldred Allen is an Inuk photographer from Rigolet, Nunatsiavut, NL, who has garnered attention for his expansive and stunningly lit scenes of landscape and wildlife in his community, whose composition and coloring elevate their everyday subject matter to the extraordinary. His artistic journey began in 2018 with the purchase of his first camera, and this self-taught photographer has been in nearly a dozen exhibitions throughout Canada in his short 6-year journey. He is represented by the Stephen Bulger Gallery of Toronto and La Guilde Art Gallery in Montreal.

Chris Dunn is currently living in Iceland as a Fulbright scholar with an NSF Arctic Research Grant focused on his project, "Applying the Environmental Humanities to Conservation Management and Policy in Iceland." A key aspect of this investigation is the ongoing conflict between conservation and renewable energy development in Iceland's Central Highland. Chris has taught at Rocky Mountain College in Montana, the Colorado School of Mines, and as a contract college instructor for the U.S. Navy in Japan and for six months on the open ocean. His PhD is in Environmental Studies from the University of Colorado-Boulder.

Erin Ggaadimits Ivalu Gingrich is a Koyukon Denaa and Iñupiaq carver, interdisciplinary artist, and scholar working and subsisting in South-Central Alaska on Dena'ina homelands. Honoring her arctic and subarctic ancestral homelands, Ivalu's work represents what has tied her and her ancestors to the North. Through carved, painted, and beaded sculpture and mask forms, Ivalu creates representations of the revered wild relatives that have provided for her, her family, and her ancestors since time immemorial. Connection to the realities of subsistence lifeways and arctic survival is vital to Ivalu's work that mirrors what keeps us fed and present in the North.

Brendan Griebel is an Arctic-based anthropologist whose work supports Inuit communities in the resurgence of traditional language and culture. His research seeks to document and revive historical Inuit lifeways—including food procurement, resource management, and consumption—so as to better navigate the impacts that climate change is having on food security and the pursuit of traditional economies. Throughout this chapter, his narrative voice is in dialogue with the artistic works of his brother Jude Griebel.

Jude Griebel is a contemporary artist whose work creates detailed figurative sculptures that visualize contemporary human entanglement with the surrounding world. In his artwork, landscapes, the species we affect, and the waste we create coalesce in vivid forms that illustrate the reach of human impact on the natural environment. Jude's combination of artwork and writing speaks to the many ways in which global consumption impacts diverse landscapes and ecologies, including the contemporary Arctic.

Mirja Hiltunen (Doctor of Art, MEd) is a professor of Art Education in the Faculty of Art and Design, University of Lapland, Rovaniemi, Finland. She has devised a performative art as part of her work in art teacher education and has been leading community-based art education projects in Lapland for over 20 years. The place-specificity, performativity, and socially engaged art are particular interests of hers. She has presented numerous international research papers and published her work in art education journals and books and art exhibitions. She is one of the leaders of the Northern Art, Community and Environment Research group (NACER) (https://nacerteam.weebly.com/).

Petri Hoppu, PhD, is a principal lecturer of dance at the Oulu University of Applied Sciences and an adjunct professor (docent) in dance studies at Tampere University. His expertise includes theory and methodology in dance anthropology and research of Skolt Saami dances, Finnish-Karelian vernacular dances, and Nordic folk dance revitalization. He has co-edited the book *Nordic Dance Spaces: Practicing and Imagining a Region* (2014) with Professor Karen Vedel and a special issue of Dance Research Journal, *In and Out of Norden: Dance and the Migratory Condition* (2020), with Professor Inger Damsholt. He has been on the boards of international dance scholars' associations and giving lectures at universities in Europe and the United States.

Maria Huhmarniemi is Vice Dean and Associate Professor at the Faculty of Arts and Design, University of Lapland, Finland, and vice lead of the Arctic Sustainable Arts and Design Network within the University of the Arctic. She is dedicated to creating activist art and pioneering arts-based methods to address environmental and societal issues in the Arctic. Huhmarniemi also carries the title of docent at the University of Eastern Finland, specializing in the field of social pedagogy. Her contributions extend to publishing research on art education and Arctic art and participating in art exhibitions, showcasing her research as an installation artist (https://orcid.org/0000-0002-3521-0679).

Esa Pekka Isomursu is a principal lecturer of media at the Oulu University of Applied Sciences, Finland, and a photographer-researcher of Northern nature who contemplates the boundaries of authenticity in landscape photography and how a photograph can convey an expanded nature experience. In his artistic research, Isomursu uses various photography techniques to examine the embodiment of nature experience and the limitations of human senses. He holds a Doctor of Technology (SW & El.Eng.) degree from the University of Oulu and is currently a doctoral student in Art at the University of Lapland, Faculty of Art and Design, Finland.

Timo Jokela is Professor of Art Education and Visual Art at the Faculty of Art and Design, University of Lapland. He is Chair of Art, Design and Culture, University of Arctic and a leader of the University of Arctic's thematic network on Arctic Sustainable Art and Design (ASAD) since 2012. From 2010 to 2019, he worked as the head of the Northern Culture Institution of Lapland University Consortium. He also worked as Visiting Professor of Art Education and Environmental Art at the University of Strathclyde, Glasgow, Scotland (2006–2011). His theoretical academic studies and art-based research focus on the relationship between Arctic cultures, arts, and nature. He has been responsible for several international, national, and regional art, nature, and culture research projects, where participatory new genre Arctic art has been studied in the contexts of decolonization, revitalization, and cultural and social sustainability. Jokela works actively as an artist, often using natural materials and the cultural heritage of the Arctic as a starting point for his community-based works and artistic projects. In 2019, he was awarded the International Society of Education through Art Edwin Ziegfield Award for his leadership in forging a new direction in art education.

Jonathan Katz is an illustrator and visual storyteller. In 2021, he received his MA in Visual Communication from the Royal College of Art, London, UK. Before moving to London, Jonathan lived in Kyoto, Japan, where he worked as a freelance illustrator. From 2019 to 2021, he was a collaborator with "Fishskinlab", a project funded by EU Horizon 2020, creating textile artworks produced on fish leather. For the past few years, Jonathan has been experimenting with his practice as a narrative illustrator in the virtual world. In his work, Jonathan is passionate about handmade techniques, particularly the essence and the concept of craftsmanship. He believes that by preserving the knowledge of traditional methods, real innovation begins. Currently, Jonathan explores how such techniques may be adapted to digital spaces and ways of making.

Korinna Korsström-Magga (MA, artist) has developed a special interest in participatory community art education in the North in order to enlighten and increase understanding between different communities, through her freelance work as an art educator. In her doctoral study (in review process) for the Faculty of Art and Design at the University of Lapland, Korsström-Magga is exploring the reindeer herders' contemporary daily living, involving the reindeer herders themselves to participate in the collection of visual research data. The project reveals their daily lives and their livelihood to other stakeholders who are branching out on their ancient reindeer pastures.

Anja Kath Lande is a certified jeweler with a diploma in jewelry design studies. Her work has been exhibited internationally, and she is devoted to both handicraft and design. She also has a master's degree in art history from the University of Tromsø, Norway. She currently holds the position of intendant at Adde Zetterquist Art Gallery. Her field of interest is wide, but environmental issues have been a red line throughout her creative work and education. She has a curious mind and allows herself to investigate interdisciplinary themes with a childlike pursuit, always eager to acquire knowledge. She is currently doing research into writing a biography about the artist Kajsa Zetterquist.

Nina Luostarinen is a DA from the University of Lapland, Faculty of Arts and Design. She currently works at Humak University of Applied Sciences as an RDI specialist and senior lecturer. She has a background in puppetry, media art, and cultural management. Both her research and artistic interests intertwine with topics of imaginary worlds, playfulness, serendipity, connection to places, and facilitating emotions that can ease attachment and empathy.

Hannah Perrine Mode (she/they) is an artist, educator, and designer working at the confluence of art and environmental science. Through making, teaching, and organizing, she harnesses creative practices to deepen human connection to Earth systems and foster intersectional approaches to climate justice. Incorporating place-based making and cross-disciplinary exchange, her artworks and installations explore geologic forces, stewardship, and memory-keeping. Hannah works extensively with environmental and outdoor organizations to develop curricula, create programming, and teach art as a method of research, experimentation, and communication. She currently serves in faculty and leadership roles with the Juneau Icefield Research Program and Inspiring Girls* Expeditions. She has previously worked with Chulengo Expeditions, the International Arctic Science Committee, Parks Canada, Alaska State Parks, Google, Vermont Studio Center, and Central Michigan University. Hannah has exhibited at museums and galleries around the country, in addition to creating large-scale permanent public artworks in the Bay Area. She is the Design Director at Storyroot, a small communication studio specializing in environmental and social impact strategy. She holds an MFA in Visual Art from Mills College and a BS in Studio Art from Skidmore College. Hannah lives in Providence, RI, and spends her summers in Alaska.

Elisa Palomino is a fashion designer, educator, and Arctic fashion anthropologist with a PhD in Sustainable Fashion and Anthropology from the University of the Arts, London. Elisa has led EU-funded projects such as Horizon 2020 FishSkin; she has been a Fulbright scholar, a Kluge fellow at the Library of Congress, a fellow at the Max Planck Institute for the History of Science, and is currently a joint fellow at Kunsthistorisches Institut in Florenz, Max-Planck-Institut, and ANAMED. She is a Research Associate at the Smithsonian Arctic Studies Center, specializing in Indigenous cultural heritage, Traditional Ecological Knowledge, Visual Anthropology, and Museum Studies. Elisa holds 25 years of experience designing for luxury fashion brands, including John Galliano and Christian Dior. As Fashion Print director at Central Saint Martins, London, and international lecturer for over a decade, she focuses on sustainability and nature-based materials.

Jessica Penney is an Inuk researcher raised in Iqaluit, Nunavut, with family ties to Rigolet, Nunatsiavut, and Carbonear, Newfoundland. Her work focuses on the lived experience of environmental change in both Nunatsiavut and Nunavut, including the impacts of climate change and environmental contaminants on Inuit food security and sovereignty. She is currently a Provost's Postdoctoral Fellow at the Dalla Lana School of Public Health at the University of Toronto, where she researches and teaches about Inuit wellbeing and Northern environmental issues.

Antonia Sohns earned her PhD in Geography from McGill University (2020). Her research and work focus on water security and climate change adaptation. She holds an MSc in Water Science, Policy and Management from the University of Oxford (2011) and a BSc in Earth Systems, Oceans track from Stanford University (2010). She has conducted environmental and social science research from the Arctic to the tropics, including for the Smithsonian Tropical Research Institute in Panama and for international NGOs.

Vera Solovyeva an Indigenous (Sakha) person from the Sakha Republic (Yakutia), Russian Federation, grew up in a remote Siberian village and now lives in the United States. She is a fellow at the Arctic Studies Center (Smithsonian Institute, DC). Her research focuses on how indigenous peoples preserve and develop their cultures and traditions in a contemporary world that is rapidly changing under the pressure of factors such as globalization and climate change. Additionally, she is interested in how indigenous peoples recover lost knowledge and traditions through the study of museum collections. Thus, starting from 2012, she co-organized multiple visits of traditional masters from the Sakha Republic to the American Museum of Natural History (AMNH) in New York City. In 2014, she co-organized the Sakha ritual "Ihiakh" and Days of Sakha culture in California. Vera Solovyeva also advocates for social justice and the Empowerment of Indigenous women. She represents Indigenous organizations from the Republic of Sakha (Yakutia) in this field at the UN Permanent Forum on Indigenous Issues. She is also a member of the Global Advisory Board of the Global Alliance of Indigenous Peoples, Gender Justice, and Peace. She believes that indigenous peoples' knowledge, ethic, and values can help the world be better for all people.

Antti Stöckell is an artist and has worked as a lecturer in art education and applied visual art at the University of Lapland, Faculty of Art and Design, Finland, since 2008. He has a degree in Fine Arts and holds a master's degree in Art Education. As a doctoral student, his research interest is in using art as a tool for processing experiences of environmental change. As an artist, he focuses on working in the field of environmental art.

Acknowledgments

This volume would not have been possible without the generous contributions, insights, and creativity of the authors whose work appears within these pages. Each has brought a distinctive perspective to the exploration of artistic witness and response in the Arctic, providing critical reflections and original artistic visions that illuminate the profound environmental changes shaping and affecting this unique region. By enabling direct engagement with the Arctic's places, stories, and unfolding transformations, they have enriched both the work of this volume and our collective understanding of the Arctic's fragile splendor. Thank you.

I am grateful to our publisher, and particularly to our editor Grace Harrison, for her support and guidance throughout this project. Routledge's commitment to fostering scholarship and disseminating new knowledge has been instrumental in bringing these diverse voices and perspectives together.

I also wish to acknowledge the influence of the Arctic itself—its landscapes, communities, and ecosystems. Observing and experiencing its transformative power has shaped the direction of this collection. The opportunity to witness firsthand the beauty and stark realities of the Arctic's environments underpins the themes and messages woven throughout the text. In reflecting on the Arctic, we are reminded of the importance of fully recognizing and valuing the environments that surround us, wherever we may live. Each place, with its defining qualities and cycles, shapes not only our lived experiences, but also our very sense of identity. By cultivating an appreciation for these diverse and ever-changing landscapes, we can deepen our connection to the world around us and foster greater care for the natural environment.

Finally, thank you to our families, friends, and communities. Without support from you all, this collection would not have been possible. Thank you.

Introduction

Antonia Sohns

The Arctic is a place of edges and convergence, where ice meets ocean, land meets sky, and cultures, human and non-human, have long intersected in a delicate balance. It is a region of dramatic change and seasonal cycles. The pulse of the planet contracts and expands in the growth and melting of sea ice and glaciers. Now those planetary cycles are being profoundly altered with the rapid onset of climate change. The Arctic warmed nearly four times faster than the rest of the world from 1979 to 2021, intensifying disruptions to ecosystems and ways of life [1].

As the impacts of climate change manifest across vast geographies, many struggle to grasp the scale of the disruptions and how they affect the people and life that inhabit these lands and waters. Similarly, other Arctic environmental challenges, such as pollution, ocean warming, or water contamination, are frequently diffuse and incremental, making them equally hard to detect and observe. As artists focus on the intimate details of a place, they observe and document the ephemeral and abrupt changes occurring around them. Artists, therefore, offer a means of witnessing, communicating, and responding to environmental changes through the creative production of aesthetic works. The word "aesthetic" comes from the Greek *aisthētikos*, "of sense perception" and *aisthanesthai*, "to perceive," underscoring the deep connection between art and sensory engagement.

Artistic responses to environmental change frequently demonstrate a strong sense of place, arising from intimate connections with specific landscapes. These connections yield nuanced insights into how climatic shifts affect health, identities, cultural practices, and bodies of knowledge. Through diverse works of art, such as music, visual art, and literature, artists interpret and depict elements of the environment so that others may become aware of those other existing realities through direct sensory experience. By fostering embodied awareness and eliciting emotional resonance, artistic works refine our perception of phenomena that might go unnoticed. In doing so, they give us the experience of being woken up to the world and encourage us to look closer, seeing things for the first time even if those realities have always existed.

The Arctic landscape is an animate, relational world where land, water, ice, sky, and their beings are kin. Leonard et al (2023) underscore that water itself,

DOI: 10.4324/9781003545880-1

like land, is a living relative, demanding an ethics of care and accountability that extends across cultural, political, and legal spheres [2]. This ontological understanding is integral to the artistic and cultural traditions that have developed in the Arctic, where stories, songs, artwork, and life do not simply depict the world but emerge from and contribute to it [4, 5]. Art serves as a means of self-reliance, sovereignty, cultural resilience, environmental stewardship, and of sharing history and memories so they are not forgotten between generations [3, 6, 7].

Language is one art form that has always been used to share knowledge and identity. The language we employ to describe landscapes tells what we see or recalls what we have seen, and conveys the history of place through its named features and stories, often revealing how we try to make "what is separate from us a part of where we are" [11]. Memories are inscribed in the land and in the names of places. Indigenous place names "are windows into the culture and history of people who have long lived there" [36]. The names may refer to a geographic feature, to a person, or to "someone or something in the legendary realm … or a resource such as fish or caribou" [36]. Through the naming of places and sharing them through stories and language, knowledge is concretized and our connection to land is remembered.

The places are written, engraved, and printed in our minds, drawn within us. "Returning to a place, experiencing its material presence, plays an active role in recollection in a profound sense, linking persons and events, situating them in a landscape" (p. 295) [12]. Those memories may return with the sight of familiar landscapes, the smells, sounds, or tastes, or through the touch of landscape beneath hand, between fingers, or under foot. Remembering the terrains of travel through bare feet, as "textures, sensations, resistances, durably imprinted memories … footnotes, born of the skin of the walker meeting the skin of the land" (p. 159) [13].

As places are changed with the loss of endemic species, permafrost thaw, ice breakup, or lands eroding into the rising sea, our ability to recall memories is challenged, and new memories are born from the transformed landscapes. The lost landscapes and the memories that inhabit them have resulted in long-term emotional impacts and existential distress. Terms like solastalgia, eco-anxiety, and climate anxiety have emerged to describe the stress of perceived, experienced, and anticipated environmental changes to existing ecosystems and landscapes [14, 15]. Ecological grief and climate grief are rising in use to describe feelings of loss related to ecological communities, vanishing forms of ecological knowledge, and the disappearance of place-specific activities, such as dog sledding or ice skating on local water bodies that no longer freeze in warming winters [9, 15, 16]. These emotions are tied to both nostalgia of place that arises as people must move away from landscapes that are no longer habitable, and the pain that is experienced when people watch a familiar landscape be "rendered unrecognizable by climate change or corporate action" [17]. Climate and its changing are thus woven into our individual and community identities, our personal and communal landscapes, and the future we will collectively inhabit.

While the psychological effects of climate and environmental changes are not fully understood, it is clear that their impacts are disproportionate and differential. Historically marginalized populations have been subjected to environmental injustices that continue to endure and affect their lived experiences. Experiences of environmental change and geographical differences, such as exposure to climate change impacts due to place, and insecurities related to land, food, or water availability, or energy access, manifest in our bodies and tell a story about the conditions of our existence [18, 19]. In water security research, for instance, scholars have documented how impacts from the lack of access to drinking water are embodied through health impacts such as physical stress of searching for additional water, or the mental stress of rationing existing water supplies [20–24]. Similarly, studies of climate change and extreme weather have revealed that stress, anxiety, depression, and fear are increasing as people experience flooding and storm surge that threaten property, communities, cultural places, and human life [25, 26].

Globally, the mental and physical impacts of climate change are diverse and nuanced. How environmental changes affect health and identity will continue to evolve, especially as its full impacts on cultural resources, historical places of subsistence and tradition, and spaces of shared memory and kinship are realized. Artistic engagement with environmental and cultural injustice has gained increasing scholarly attention as a powerful strategy for community healing and empowerment, particularly in regions like the Arctic that are facing the acute impacts of climate change [27].

Within this context, art, in its many forms, becomes a powerful means of direct immersion in a place to cultivate deeper emotional and somatic connections to the environment, often prompting a sense of urgency and stewardship [8]. Artistic approaches to documenting and interpreting Arctic change are increasingly recognized as powerful complements to scientific inquiry, offering nuanced insights into the social, cultural, and ecological dimensions of environmental change [8, 27]. Recent scholarship highlights how creative engagement, ranging from visual narratives to interactive installations, can contextualize climate data within lived experiences, thereby amplifying public awareness and emotional resonance [9, 30]. Through collaborative creativity, participatory art, and interactive play, people can restore ecological and cultural connections and foster empathy and mutual responsibility between communities and their environment [28, 29].

Such integrative practices not only help to witness and respond to environmental change, but also revitalize local ecocultures by foregrounding traditional knowledge and perspectives [5]. Visual arts, such as photography, are potent mediums for community-based monitoring and advocacy, enabling the documentation of on-the-ground environmental change, shaping of perception, and the expression of emotional and cultural responses to such shifts [31]. By engaging with Indigenous languages, material cultures, and epistemologies, creative projects emphasize the significance of ancestral knowledges in adapting to new climatic realities, underscoring the critical role of art in witnessing, shaping, and sustaining Arctic futures [8]. Moreover, integrating culturally significant materials and participatory creative processes can amplify community voices, particularly those of Indigenous youth and Elders, by linking personal expressions of environmental change with broader dialogues on relationships with the land and waters, and heritage [27]. In turn, artistic interventions can catalyze conversations about community values, political issues, and ecological entanglements [9, 10, 34, 37].

As the trend of languages globally continues to use non-descript words and terms, such as lake, bird, flower, instead of the specific species or geographic name, it reveals societies' growing lack of familiarity with nature [11, 26]. This disconnection between people and nature may mean people do not see the changes or do not have the words to describe the impacts and therefore may be unmotivated to protect the environment from continued degradation [32, 33]. Art offers a means of joining us to nature, through language and our senses, providing a closer association to landscapes and experience at this critical time of rapid change and distancing from the environment.

Through creative, embodied modes of documentation and response, the approaches in this collection reinforce the notion that an inclusive, dialogic engagement with Northern environments is pivotal not only for understanding environmental change, but also for imagining and enacting just and sustainable futures.

Why This Collection

This collection arises from a shared sense of urgency and possibility: the urgency to address accelerating environmental changes across the Arctic, and the possibility that art—in its many forms—can help us to see, feel, and respond to these transformations. This collection brings together Indigenous and non-Indigenous voices to explore how scholars and artists have combined artistic enterprise and science to convey the intimate manifestations of environmental change through artistic and sensory experiences [35]. This collection examines interdisciplinary and intercultural methodologies that combine artistic research, community engagement, and place-based inquiry. The contributors are artists, scientists, anthropologists, and many other identities of scholars and practitioners and range from early career researchers with unfolding research and artistic work before them to senior scholars and artists who have contributed richly over the years to their fields. The voices in this collection come together not as a singular narrative, but as a multi-sensory, interdisciplinary, and intergenerational conversation between Indigenous and non-Indigenous authors.

This collection is divided into three parts. The first part of the collection explores art as witness. It brings together artists, scholars, and practitioners who document and interpret Arctic change through artistic perception, situated experience, and visual storytelling. Through creative, community-centered practices, art emerges as both a means of witnessing environmental transformations and a conduit for forging profound relational bonds that propel meaningful, place-based responses to injustice. Maria Huhmarniemi and Anja Kath

Lande describe their Observation of Change project in Junkerdal National Park that blends art and biological science to document ecological changes, enhance public understanding of biodiversity, and contribute to the revitalization of Arctic ecocultures. The insights gained from this collaboration inform future educational practices and contribute to a broader global discourse on ecological and multispecies approaches in contemporary art.

Further exploring art's power as a tool for witnessing, through a collaborative project with Rigolet residents, Jessica Penney and Eldred Allen demonstrate how community photography serves as both a visual record and a means of expressing the lived experience of climate change and industrial impacts, such as contamination in Labrador. Penney and Allen's work underscores the emotional toll of environmental uncertainty and the role of visual storytelling in asserting agency and resilience.

Chris Dunn presents two multimedia projects, Raven's Light and Sensing Ice, that focus on Indigenous material culture and sensorial experiences of ice. Raven's Light explores the material culture of Indigenous peoples within Arctic Alaska and creates a collage. Sensing Ice centers on sensorial encounters with ice paired and juxtaposed with scientific renderings. These projects create a dialogue between artistic and scientific renderings of Arctic change and underscore the human relationship and interaction with material surroundings, whether as sacred landscape or phenomenological witnessing of the beauty and wonder of nature's forms, particularly amid industrial-technological change.

Further exploring art as witness, Erin Gingrich discusses Iñupiaq carving and weather through the project Silamiñ. Locating many key concepts through wayfinding processes such as research, creation, and learning by doing. Gingrich explores the recognition of carving as tangible discourse for addressing the climate in Nunaat and reasserting Iñupiaq language and cosmologies by representing weather as breath. Exploring relationships between weather, the breath, and carving as ancestral discourse, Gingrich underscores the importance of Indigenous language and epistemology to adaptation and resilience to climatic change.

The second part of this collection is centered on art as a response. Part II focuses on how art is mobilized in response to environmental and cultural injustice. Together, these chapters consider how artistic practices may enable direct, physical immersion in a particular geography or through artistic representations that evoke its essence, catalyze empathy and urgency, and encourage care and responsibility for the landscapes we inhabit or envision. Capacity to perceive becomes the wellspring from which meaningful responses to environmental challenges arise.

Petri Hoppu and Esa Pekka Isomursu examine how people perceive Northern landscapes through nature photography and develop ideas of environmental authenticity. Nature photography can evoke a multisensory and immersive experience influenced by factors such as the viewer's prior experiences, image objects, and composition. Drawing on Groys's interpretation of Benjamin's concept of aura, Hoppu and Isomursu discuss how documentary-style landscape photographs can evoke feelings of uniqueness and authenticity and shape viewers' perceptions of Northern landscapes. Hoppu and Isomursu discuss new ways of depicting the Northern landscape in photographs and visual arts to evoke multisensory engagement and how visual culture influences perceptions of environmental authenticity.

The following chapter by Mirja Hiltunen and Korinna Korsström-Magga presents participatory art with Sámi youth and young people from diverse Arctic backgrounds. These youths expressed their hopes, concerns, and aspirations for a sustainable future by carving words, symbols, and messages onto reindeer antlers. Antlers are a culturally significant material in Sámi handicraft (duodji), serving as a medium to foster dialogue on the impact of climate change on Arctic livelihoods, culture, and nature. Their collective artwork is now installed permanently in front of the Sámi Parliament, Sajos, in Inari, Finland. The installation not only represents the voices of the only Indigenous people in Europe who are at risk due to climate change, but also serves as a symbol of their agency and resilience.

Further exploring how art can be used as a response to environmental change, Timo Jokela describes his project Kuerin Matka (Kueri's Journey), developed in response to a proposed iron ore mine near the Pallas-Yllästunturi National Park in Northern Finland. Using arts-based action research, Jokela created a dialogic space for local concerns about environmental conflict and cultural change. In this chapter, Jokela discusses the relationship and encounter between Western contemporary art practices and Northern ways of knowing nature. Vera Solovyeva also investigates differences between Western and Indigenous art in her chapter. Solovyeva's chapter describes how artists and traditional masters in the Republic of Sakha (Yakutia) respond to the local impacts of climate change, and how they use artistic expression to foster emotional connection, increase awareness, and motivate action.

Brendan Griebel and Jude Griebel examine ways in which global consumption impacts diverse landscapes and ecologies, including the contemporary Arctic, through a dialogue between historical Inuit lifeways and detailed figurative sculpture that visualizes contemporary entanglement with the surrounding environment. Using art and anthropological research, Griebel and Griebel discuss the reach of global consumption and its impact on the ecosystems and communities in the Arctic.

The third part of the collection centers on artistic practices that may deepen connection or reconnect people or communities with the environment and cultural heritage. In their chapter, Antti Stöckell and Nina Luostarinen examine how art-based play and engagement can serve as tools for ecological and psychological restoration. In a comprehensive discussion, Stöckell and Luostarinen discuss how, by combining nature, culture, and art, it is possible to address difficult questions of environmental change caused by human activity through the content of the play. Stöckell and Luostarinen highlight how collaborative creativity can restore connections to the land and to each other and encourage new perspectives on environmental stewardship and sustainability.

Focusing on how creative practices can deepen human connection to Earth systems and enhance engagement with polar science, Hannah Perrine Mode describes her integrative work with the Juneau Icefield Research Program. Mode uses artistic methods to bridge science, pedagogy, and environmental stewardship. Mode's narrative addresses the challenges in communicating climate science and glacier retreat and emphasizes the crucial role of interdisciplinary and collaborative work in addressing global climate change. Mode's artistic and scientific practices both document and respond to the environmental changes being experienced in the North.

Elisa Palomino and Jonathan Katz explore how technology-enabled access to cultural heritage, such as digital archives and virtual reconstructions of clothing made from environmental materials, can open new possibilities for reclaiming colonially displaced knowledge and ensure that Arctic peoples maintain intellectual sovereignty over their histories and future trajectories. Palomino and Katz discuss challenges Arctic Indigenous communities face in accessing their cultural heritage, as artifacts collected by explorers are housed in distant museums. This disconnection causes intellectual and aesthetic losses, exacerbated by geographical distance and prohibitive travel costs. Through the lens of fish skin garments, their chapter examines the possibilities of 3D technologies for providing Arctic Indigenous communities with improved access to cultural heritage using AI and digital tools.

Drawing on nearly 50 years of artistic practice, Timo Jokela discusses rivers as dynamic ecosystems and cultural lifelines, using diffractive reading to bridge Western and Indigenous knowledge systems. Rivers are strongly influenced by human-induced changes, such as natural resource exploitation or climate change. Jokela details how rivers, particularly the Ounas River, are sites of cultural and ecological memory. Jokela's chapter importantly considers the potential of art to act as a lever for Arctic sustainability transformation.

Reading This Collection

This collection proffers an exploration of art as a dynamic means of witnessing, interpreting, and responding to environmental changes in the Arctic. Together, this collection reveals the multiple roles art plays: bearing witness to environmental transformation, amplifying the lived experiences of climate disruption, and serving as a medium of healing and connection to community, knowledge, and tradition. Whether through play-based environmental engagement, multimedia storytelling, digital reconnections with material culture, or place-specific artistic interventions, the chapters presented here challenge conventional representations of Arctic change. They highlight how creative practices make visible slow and rapid environmental shifts, and offer alternative ways of perception and response.

Rather than framing the Arctic as a place of loss alone, this collection underscores its resilience, creativity, and ongoing knowledge-making. In the face of accelerating climate change and the pressures of resource extraction, industry, and shipping, these artistic practices reveal the Arctic not only as a site of environmental precarity, but also as a space of agency, continuity, and profound interconnection. Art invites us to recognize the deep entanglements between human and more-than-human worlds and to imagine futures that honor and sustain them.

References

1. Rantanen, M., Karpechko, A.Y., Lipponen, A. et al. (2022). The Arctic has warmed nearly four times faster than the globe since 1979. *Commun Earth Environ*. 3: 168. doi:10.1038/s43247-022-00498-3
2. Leonard, K., David-Chavez, D., Smiles, D., Jennings, L., Anolani Alegado, R., Tsinnajinnie, L., Manitowabi, J., Arsenault, R., Begay, R.L., Kagawa-Viviani, A., Davis, D.D., van Uitregt, V., Pichette, H., Liboiron, M., Moggridge, B., Russo Carroll, S., Tsosie, R.L., Gomez, A. (2023). Water back: A review centering rematriation and Indigenous Water research sovereignty. *Water Altern*. 16(2): 374–428.
3. Igloliorte, H. (2009) *Response, Responsibility, and Renewal. Canada's Truth and Reconciliation Journey*. Aboriginal Healing Foundation Research Series. Aboriginal Healing Foundation. ISBN 978-1-897285-72-5. https://ehprnh2mwo3.exactdn.com/wp-content/uploads/2021/01/trc2.pdf
4. Cruikshank, J. (2005). *Do Glaciers Listen? Local Knowledge, Colonial Encounters, and Social Imagination*. UBC Press.
5. Whyte, K.P. (2017). Indigenous climate change studies: Indigenizing futures, decolonizing the Anthropocene. *Eng Lang Notes*. 55(1–2): 153–62.
6. Cunneen, C. (2017). *Visual Power and Sovereignty: Indigenous Art and Colonialism*. Routledge International Handbook of Visual Criminology. Routledge. ISBN 9781315713281
7. von Harringa, C. (2023). Ecopoetics, Decolonizing Aesthetics, and the Fluid Spaces of Co-Existence in Contemporary Circumpolar Indigenous Art and Exhibitionary Practice. PhD thesis, Concordia University. https://spectrum.library.concordia.ca/id/eprint/992890/
8. Tschakert, P., Ellis, N., Anderson, C., Kelly, A., Obeng, J., Robinson, C. (2019). One thousand ways to experience loss: A systematic analysis of climate-related intangible harm from around the world. *Global Environ Change*. 55: 58–72.
9. Cunsolo, A., Ellis, N.R. (2018). Ecological grief as a mental health response to climate change–related loss. *Nat Clim Change*, 8(4): 275–281.
10. Pihkala, P. (2020). Anxiety and the ecological crisis: An analysis of eco-anxiety and climate anxiety. *Sustainability*, 12(19): 7836.
11. Lopez, B.H., Gwartney, D., editors. *Home Ground: Language for an American Landscape*. Trinity University Press; 2013. p. 672.
12. Tilley, C., Cameron-Daum, K. (2017). Anthropology of Landscape [Internet]. Anthropology of Landscape. UCL Press; [cited 2022 Aug 9]. 346 p. Available from: https://www.jstor.org/stable/j.ctt1mtz542
13. Macfarlane, R. (2013). The Old Ways [Internet]. Penguin Books; [cited 2022 Aug 9]. 448 p. Available from: https://www.penguinrandomhouse.com/books/305206/the-old-ways-by-robert-macfarlane/
14. Albrecht, G., Sartore, G.M., Connor, L., Higginbotham, N., Freeman, S., Kelly, B., et al. (2007). Solastalgia: The distress caused by environmental change. *Australas Psychiatry*. 15(SUPPL. 1). Available from: https://pubmed.ncbi.nlm.nih.gov/18027145/
15. Cunsolo, A., Harper, S.L., Minor, K., Hayes, K., Williams, K.G., Howard, C. (2020). Ecological grief and anxiety: The start of a healthy response to climate change? *Lancet Planet Heal*. 4(7):e261–3. Available from: http://www.thelancet.com/article/S2542519620301443/fulltext

16. Macfarlane, R. (2019). *Underland: A Deep Time Journey*. W.W. Norton & Company. p. 496.
17. Powell, M., Boyne, G., Ashworth, R. (2001). Towards a geography of people poverty and place poverty. *Policy Polit*. 29(3): 243–58.
18. Krieger, N. (2005). Embodiment: A conceptual glossary for epidemiology. *J Epidemiol Community Health*. 59(5): 350. Available from: /pmc/articles/PMC1733093/?report=abstract
19. Eichelberger, L. (2018). Household water insecurity and its cultural dimensions: preliminary results from Newtok, Alaska. *Environ Sci Pollut Res*. 25(33): 32938–32951. doi: 10.1007/s11356-017-9432-4
20. Mushavi, R.C., Burns, B.F.O., Kakuhikire, B., Owembabazi, M., Vořechovská, D., McDonough, A.Q., et al. (2019). "When you have no water, it means you have no peace": A mixed-method, whole-population study of water insecurity and depression in rural Uganda. *Soc Sci Med*. 245:112561. doi: 10.1016/j.socscimed.2019.112561
21. Marino, E., White, D., Schweitzer, P., Chambers, M., Wisniewski, J. (2009). Drinking water in Northwestern Alaska: Using or not using centralized water systems in two rural communities. *Arctic*. 62(1): 75–82. Available from: www.jstor.org/stable/40513266
22. Eichelberger, L. (2017). Household water insecurity and its cultural dimensions: Preliminary results from Newtok, Alaska. *Environ Sci Pollut Res*. 25: 32938–32951. 10.1007/s11356-017-9432-4
23. Rosinger, A.Y., Young, S.L. (2020). The toll of household water insecurity on health and human biology: Current understandings and future directions. *Wiley Interdiscip Rev Water*. 7(6):e1468. Available from: https://onlinelibrary.wiley.com/doi/full/10.1002/wat2.1468
24. Rosinger, A.Y., Bethancourt, H.J., Young, S.L., Schultz, A.F. (2021). The embodiment of water insecurity: Injuries and chronic stress in lowland Bolivia. *Soc Sci Med*. 291: 114490.
25. Sarkar, A., Hanrahan, M., Hudson, A. (2015). Water insecurity in Canadian Indigenous communities: Some inconvenient truths. *Rural Remote Health*. 15(4): 3354.
26. Stambach, A.E., Kwayu, A.C. (2021). Witness to a passing. *J Ethnogr Theory*. 11(2): 412–27. Available from: https://www.journals.uchicago.edu/doi/10.1086/716216
27. Galafassi, D., Kagan, S., Milkoreit, M., Heras, M., Bilodeau, C., Juárez Bourke, S., Merrie, A. (2018). 'Raising the temperature': The arts in a warming planet. *Curr Opin Environ Sustain*. 31: 71–79.
28. Ingold, T. (2000). *The Perception of the Environment: Essays on Livelihood, Dwelling and Skill*. Routledge.
29. Kimmerer, R.W. (2013). *Braiding Sweetgrass: Indigenous Wisdom, Scientific Knowledge, and the Teachings of Plants*. Milkweed Editions.
30. O'Neill, S.J., Smith, N. (2014). Climate change and visual imagery. *Wiley Interdiscip Rev Clim Change*. 5(1): 73–87.
31. Crate, S.A., Nuttall, M. (Eds.). (2016). *Anthropology and Climate Change: From Actions to Transformations*. Routledge.
32. Kesebir, S., Kesebir, P. (2017). How Modern Life Became Disconnected from Nature [Internet]. Greater Good Magazine. Available from: https://greatergood.berkeley.edu/article/item/how_modern_life_became_disconnected_from_nature
33. Cervinka, R., Röderer, K., Hefler, E. (2012). Are nature lovers happy? On various indicators of well-being and connectedness with nature. *J Health Psychol*. 17(3): 379–88.
34. Martineau, J. (2015). Creative Combat: Indigenous Art, Resurgence, and Decolonization (Doctoral dissertation). https://dspace.library.uvic.ca/bitstream/1828/6702/1/Martineau_Jarrett_PhD_2015.pdf

35. Landau, F., Toland, A., Balkin, A., Husberg, H., Cerpina, Z., Stenslie, S., et al. (2021). Towards a sensory politics of the Anthropocene: Exploring activist-artistic approaches to politicizing air pollution. *Environ Plan C Polit Sp.* 40(3): 629–47. Available from: https://journals-sagepub-com.ezproxy.neu.edu/doi/full/10.1177/23996544209816 01
36. Canada, N.R. (2025, January 16). *Indigenous Place Names.* Natural Resources Canada. https://natural-resources.canada.ca/maps-tools-publications/maps/geographical-names-canada/indigenous-place-names
37. Martineau, J., Ritskes, E. (2014). Fugitive indigeneity: Reclaiming the terrain of decolonial struggle through Indigenous art. *Decolonization: Indigeneity Educ Soc.* 3(1):I–XII.

Part I

Art as Witness

1 Observation of Change as a New Genre Arctic Art

Maria Huhmarniemi and Anja Kath Lande

1.1 Introduction

While scientific and technological expertise is crucial, successful ecological restoration also requires local knowledge, community support, and morally grounded practices [1]. There is a growing recognition of the value of collaboration between the arts and sciences in addressing complex environmental, social, and cultural challenges and supporting nature conservation [2]. Intensive, place-based, and transdisciplinary art–science programs have the unique ability to create narratives that effectively communicate the patterns and processes of nature and human–environment interactions [2]. Artists, through their use of aesthetics and imaginative language, play a key role in fostering empathy for ecological restoration [3]. Their work can influence public understanding and concern about species loss [4] and initiate value shifts toward sustainability and a deeper respect for natural ecosystems [5].

In Norway and elsewhere in Europe, the planting of spruce trees in deciduous forests has affected forest ecosystems, worsening species loss at the age of the sixth major extinction event on Earth [6, 7]. In the Junkerdalsura Nature Reserve in Nordland, Norway, native mountain birch forests are threatened by the spread of spruce that were planted in parts of the reserve between the 1920s and 1960s when the Norwegian government tried to promote forestry. Nowadays, spruce plantations take space from many native species and spread, sometimes invasively, outside the tree plantations.

Ecological restoration may restore mountain birch ecosystems damaged by plantations (Figure 1.1). Logging was done in two spruce plantations in 2018 and 2019, and tree "veteranization," where chainsaws were used to cut, decorticate, weaken, and prematurely age trees, was applied to a spruce plantation in 2021 [7, 8]. The cut and veteranized spruces are left to rot, aiding endangered insects that need rotting wood as a habitat. Biodiversity research will continue until 2028 and map the most effective restoration methods to inform others battling expanding spruce plantations [6–8]. While ecological restoration aims to help natural biodiversity, it also triggers questions about environmental ethics, cultural sustainability, and the human role and responsibility in controlling and aiding more-than-human nature.

DOI: 10.4324/9781003545880-3

Figure 1.1 The native birch forest is on the *left* and the spruce plantation is on the *right*. The spruce trees have been harmed as part of the ecological restoration process. Photograph by Birgitta Linhart, 2023.

Observation of Change (OOC) is an art and research-creation project (2023–2025) that brings artists into collaboration with a nature restoration project in the Junkerdal National Park area to witness environmental change by spruce veteranization and logging, and to reflect on and communicate nature conservation through art.

A group of artists runs a series of art residencies in the Junkerdal National Park area in partnership with national park managers and the Adde Zetterquist Art Gallery, located next to the visitor center. Artists Birgitta Linhart and Maria Huhmarniemi initiated this project with art curator Anja Kath Lande, national park manager Johan Rova, and ecologist Rannveig Margrete Jacobsen. The collaboration includes discussions with researchers from the Norwegian Institute of Nature Research and the Norwegian University of Life Sciences, conducting biological research on biodiversity at the conservation sites as well as on reference forest areas.

In this chapter, we present and discuss art created as part of the OOC project in the framework of New Genre Arctic Art [9], a concept defined to describe contemporary artistic interventions and performances that include a political agenda to enhance sustainability in the changing Arctic. New Genre Arctic Art aims to foster the decolonization of the Arctic from Western values and definitions of art and knowledge [10]. We have been actively developing the OOC as an art intervention to reflect the ethics of ecological restoration and communicate research on biodiversity and veteranization. The research on OOC follows principles of art-based action research (ABAR) [10], in which researchers are active members of the reflective process. In this chapter, we discuss how artworks created in the OOC document nature, conceptualize its changes, and widen the general understanding of environmental change through nature conservation in the Arctic. The research question is how to support the decolonization of nature and art in the context of an Arctic national park.

As mentioned above, the biological restoration sites that are the focus of this project are situated in Junkerdalsura Nature Reserve. The nature reserve is an extension of the Junkerdal National Park, and its aim is to protect a species-rich mountain birch forest on calcareous soils in the Saltdal area, just north of the Arctic Circle. The national park and the nature reserve (together with several other protected areas) are managed by the Midtre Nordland National Park Board. At Storjord, next to the protected areas, there is also a visitor center that holds an exhibition showing the natural and cultural heritage values of the areas managed by the national park board. Although we realize that—from an administrative and legal perspective—it is important to distinguish between different kinds of protected areas, we argue that questions regarding the decolonization of (protected) nature can be seen as independent of an area's conservation status or administrative boundaries. For this reason, we have decided to refer to all OOC actions as performed in the "Junkerdal National Park area," regardless of whether they actually occur in Junkerdalsura Nature Reserve, Junkerdal National Park, at the visitor center, or at any other sites that are relevant for the questions addressed.

The OOC is an ongoing project. Six artists, one art-science collective, and an artist-curator participate in the OOC. For this chapter, we have selected art productions from four team members (including the co-authors) that resonate with the book's theme: witnessing changes in nature and clarifying the artist's role in communicating biodiversity research and renewing ecocultural traditions. As artist-researchers, we are researching a process that we are initiating, facilitating, and contributing to. The first author, Huhmarniemi, is fostering a pedagogical turn of art-science collaborations to enhance dialogue in environmental conflicts involving biodiversity issues and natural resource extraction [11–13]. In the OOC, her role is as an artist, one of the coordinators, and a researcher. Lande is a jeweler, artist, art historian, and curator. She holds the position of intendant at Adde Zetterquist Art Gallery, the key partner of the OOC. She is engaged in learning about spruce-based ecocultures and introducing them to the rest of the team. Both Huhmarniemi and Lande are revitalizing local ecocultures in the Arctic by crafting from natural materials. Lande is also involved in fishing and hunting; Huhmarniemi picks edible berries and dye plants and discusses these cultures in her research [14–16]. Thus, this chapter continues long-term actions to revitalize Arctic ecocultures as members of local communities.

The OOC is a collaborative project with partners from the University of Lapland, Nord University, and the Norwegian University of Life Sciences. Experience from this art-science collaboration is expected to be beneficial for higher education. The research aligns with a global turn toward post-humanist, ecological, and multispecies approaches in contemporary art and university education [17]. The focus on the Arctic emerges from our existing relationships with the Arctic and contributes to artists' and art educators' training in the Arctic, while we communicate process learning and research in the Arctic Sustainable Arts and Design network that identifies and shares contemporary and innovative practices in teaching, learning, research, and knowledge exchange in the fields of art, design, and visual culture education.

1.2 New Genre Arctic Art for Decolonization

Artists have long used biological motifs in their work. Since the 1700s, artists have illustrated plant and animal species to understand and describe their structures. Today, artists may aim to visualize and communicate biological research data, research methods, and results evocatively [11]. Since the 2000s, artistic and art-based research has gained acceptance in academia, increasing interest in diverse collaborations between artists and scientists. Art is now seen as a valuable approach to research, aiding in research strategies, methods, and communication of results [18].

In the field of biology, for instance, Ballengée [19] advocates for transdisciplinary art practices and participatory biology programs to enhance public understanding of ecology. He facilitates eco-actions to connect communities with specific ecosystems, engaging non-specialists through experiential field trips, workshops, and biological research. Polli [20], a media artist and art scholar, focuses on weather conditions and explores how media art can increase public engagement with weather and climate science. Her work aims to foster collaboration between media art and climate science, providing public access to weather and climate data. These examples highlight the growing field of art and science collaborations, emphasizing participatory methods.

Demos [21] provides numerous examples of artworks that emerge from and respond to environmental and local struggles in his book "Decolonizing Nature: Contemporary Art and the Politics of Ecology." Decolonizing nature as a concept refers to addressing and dismantling the colonial legacies that have shaped our understanding, management, and interaction with the natural world. The historical practice of planting invasive spruce in native birch forests is now viewed as a colonial act. Art that focuses on ecological restoration and communicates its necessity and motivations plays a role in fostering this process of decolonization.

Art-science expeditions in the Arctic have been shadowed by cultural colonialism. Anthropologists and other researchers have been accompanied by artists and painters who have romanticized the Arctic landscape, depicting it as cold and empty [22, 23]. This stereotypical image continued in contemporary climate change-themed art, which includes modern imagery of ice and polar bears [24, 25]. Some of these colonially constructed art and sciences collaborations, from the Western art world to the Arctic, have resulted in beautiful representations of snow and ice, artworks that aim to document or communicate the effects of climate change, and conceptual art referring to scientific research methods [26]. Furthermore, in the 21st century, the Arctic region has sparked the interest of international artists to examine the changing relationships between art, media, and aesthetics in the Arctic [26, 27].

While Western art-science collaborations have often been colonial toward local cultures of the Arctic, a new trend of Arctic art is emerging that is based on insights from the region's own cultures. This trend differentiates itself from the canon of Western art history. Arctic art typically combines contemporary art and crafting with natural materials, drawing from the local and traditional knowledge of communities that still live close to nature and maintain their ecocultural heritage [14, 28].

Arctic art reflects, continues, and renews the cultural heritage of Arctic regions. It is inherently political, drawing attention to environmental concerns, cultural colonialism, and the justice of the green shift [15]. The concept of "New Genre Arctic Art" describes contemporary art, performances, and media productions that cover Arctic themes such as resource politics, nature conservation, and sustainability, emphasizing their political and participatory focus [9]. Artists are expected to empower community members to participate in discussions on resource politics and nature conservation, inform and educate global audiences, and continue ecocultural connectedness in ways that align with contemporary knowledge on climate crises and biodiversity loss [9].

1.3 Organization of OOC and Research Methodology

The ABAR [29] approach is used in the OOC to support artists' expression in communicating the reasons and results of nature conservation and its biodiversity impacts. Action research allows practitioners, such as artists, curators, and art facilitators, to investigate practical situations. ABAR aligns with reflective action research, which supports and develops creative approaches in disciplines other than art, such as education and social sciences [30]. The OOC aims to enhance art-science collaboration and networking in the Nordic Arctic, use art to document and demonstrate changes in nature as well as their reasons, improve comprehension of human cooperation with more-than-human nature, use art to tackle the complex issues related to environmental ethics, and produce a touring exhibition. Process and its results are documented and used as research data to analyze and evaluate the potential strengths and weaknesses of the project.

The cyclicity of OOC means several artistic productions based on the project's theme. For this chapter, we have selected artworks for interpretation and discussion to highlight the project's aims. ABAR is always a learning process for practitioners, organizations, and communities. It is similar to transformative action research, which emphasizes that the researcher initiates change processes and, through reflecting on their actions, also transforms themselves, for example, by adapting their practices to the era of ecological crises [31]. Bradbury [31] highlights action research as sustainability transformations that help to respond to climate change and achieve local and global sustainability goals. In the OOC, artists were challenged to learn about biodiversity, nature restoration, and art-science collaboration. We reflect on the project and its artistic results to answer the research question of how to support the decolonization of nature and art in the context of a national park in the Arctic.

The OOC team of artists, the curator, and the national park manager targeted the reasons for integrating art into the nature conservation project, the set aims, and the designed actions. The plan included community engagement through exhibitions and programs for school students and research publications.

A grant from the Nordic Culture Foundation made the project possible. More artists were invited to participate. Artists have spent three art residency periods in the Junkerdal National Park area, each lasting a week. They stayed in a cabin next to the national park and the Adde Zetterquist Art Gallery. The program during those weeks included guided walks in nature, presentations by scientists on biodiversity and nature conservation, artistic work, reflective discussions, and environmental art workshops with school students. Artists developed artistic concepts for their productions in discussions with the project team. Discussions were held throughout the process among the whole team and in smaller groups.

1.3.1 Witnessing and Representing Change

Isomursu questions when and how nature photography expresses real experiences of nature [32, 33]. He considered how modern cameras can create an "unreal reality" by depicting landscapes in the dark, at the macro level, and at different wavelengths of light.

One of Isomursu's projects in OOC involved documenting a spruce plantation in the mountain landscape from a distance. He has used standard and modified camera technology as well as thermal imaging to capture the spruce field in normal, infrared, and ultraviolet light. Isomursu used various camera technologies to examine how the "real world" could look without the limitations of human perception. Thus, he expanded human understanding of nature, photographs as documents, and the limitations of human observation skills and capacities. Camera technology acts as a collaborator in his artistic practice. The outcomes of his ABAR showcase nature photographs that advance post-human observations of nature and its changes. The series of photos from ecological conservation in Junkerdal invite the viewer to consider how the spruce field could be seen in many different ways: not only through the use of different camera technologies, but also through the eyes of animals and insects, which observe their surroundings differently than humans.

1.4 Discussion

1.4.1 Reflections on Conquest

In the first residency, August 2023, Huhmarniemi took a series of photos considering the themes of conquest, power dynamics, and control of more-than-human nature. She engaged in reflecting on the colonization and decolonization of nature through the constant pursuit of a growing economy and science. The colonization of women and nature was linked in her photographs, which aimed to invite reflection on the Western view of women being close to nature and the body and men being associated with higher knowledge and science. When arriving at the residency, Huhmarniemi was still hesitating about whether ecological restoration was just one way to continue conquest and was considering why spruce could not expand widely as free people.

Linhart also worked with the theme of conquest and showed a solo exhibition, "Proud Trophies and the Insatiable Hunger," at the Adde Zetterquist Art Gallery in the summer of 2024, considering humans' power in conquering and controlling more-than-human nature.

The installation "Skalp" measures about 2.5 m in height, 2 m in width, and 3.5 m in length (Figure 1.2). Linhart made it from fireweed, cotton thread, and spruce needles collected in the spruce plantation in Junkerdalsura Nature Reserve. The top hat, where all the cotton thread springs out, is formed from fireweed and, when seen from the back, resembles a wig with long hair floating on the floor. From the side, the installation has an organic, lively shape, like a waterfall or a whale. From a distance, it looks light and frail, but when standing close, it gives a hint of seriousness. With a slight curve, the stake from the bottom of the wall bends out into the room as if carrying a heavy weight, pointing its tip through the top of the head-shaped form. The cotton threads have spruce needles glued on, resembling barbed wire, and are still neatly arranged in a heart shape on the floor, making it look brushed and untangled. When entering the exhibition room, one must tread carefully; if unaware, one might damage it by stepping on the threads, causing the spruce needles to fall off. On the floor, spruce needles are placed under the top of the scalp, creating a reference to a pool of blood.

Linhart made *Skalp* from materials found in nature and other natural materials. The fireweed (*Chamaenerion angustifolium*) is a native plant, often colonizing roadsides and other infrastructural wastelands, bearing an impressive magenta-colored inflorescence, and spreading its feather-light seeds with the wind. It was collected by the artist and was used as a soft, cushion-like base for attaching the hair-like cotton threads. The collected seeds will not bear any fruit. The spruce needles that fall to the ground in the forest will not decompose and become energy for the roots of the surrounding trees—a representation of a complete stop in the ecosystem.

Skalp as a work can also represent the frailty of the earth, as every little touch will change its original shape and the spruce needles might fall off, leaving the cotton threads bare. As the ongoing climate change makes extracting natural resources in the Arctic previously hidden under sea ice and permafrost possible, *Skalp* becomes an image of this frailty—of man's consequential game with nature. Linhart's artwork is an example of New Genre Arctic art: it combines crafting with political, local, and global themes. The installation becomes an artistic activism addressing the political and societal sense of time, questioning the time of our existence.

1.4.2 Revitalization of Spruce-Based Ecoculture

Like many other Scandinavian people, Lande grew up in an ecoculture that included forestry. For example, planting spruce trees was a summer job for many students, who were paid based on the number of trees they planted. At the time, this was considered an investment for the future, providing sustainable materials needed for building houses. As a young girl in the 1980s, Lande helped her father plant spruce trees. While working in the landscape, her father would tell her who had planted the trees that had been chopped down the previous winter. This made her wonder if, in her lifetime, she would see the trees she planted become useful. Nowadays, many spruce plantations are not suitable for contemporary extraction methods, such as the one in Junkerdalsura. These areas are not only in protected areas but also in terrains unreachable by modern forestry standards, where a horse, sleigh, or helicopter might be the only solution for transporting trees.

Lande considered how looking to traditional ecological knowledge and traditional uses of spruce could slow the spread of spruce and still provide a resource for contemporary use. She found inspiration for spruce root weaving from the Tlingit of southeast Alaska [34]. Root binding and weaving were well known by Indigenous peoples of the Arctic [35], who learned to use what nature provides without extracting more than needed and destroying ecosystems—traditional ecological knowledge. This knowledge was applied not only for practical uses but also for ornaments and jewelry. A neatly and finely woven spruce root container would traditionally be used as a water cup or for cooking. The exterior would be rubbed with a bear's tooth to spread the fibers, making it dense [36]. Harvesting the roots is a time-consuming job, and traditionally the aim was not to harm the growth of the trees by damaging the roots.

During May 2023, Lande harvested spruce roots, learning from YouTube videos and realizing that root crafts require knowledge gained from practice. She studied online sources and talked with skilled root binders, discovering that few people continue to practice spruce root weaving. Determined to master the techniques, Lande put significant effort into learning weaving. Living by the seashore, Lande collected polypropylene ropes that had washed up on the beach—waste mostly from fish farms and fishing boats over a long period of time. In August 2023, Lande conducted a workshop for OOC members, introducing the basics of spruce root crafting and weaving techniques, and using roots but also waste polypropylene strings, which are much easier for beginners to handle and bind (Figure 1.2).

Harvesting spruce tips is another way of controlling tree growth without harming trees, a practice Lande derived from contemporary ecoculture. The tips are picked in the spring when they are fresh and light green, just as they emerge from the brown caps. By boiling the tips with water, sugar, and syrup, one can make soft drinks, beer, and wine. In May 2024, Lande picked spruce tips and experimented with different uses for this resource. She made one liter of spruce tip syrup and one soft drink concentrate, enabling OOC project members to try them.

Spruce root binding and spruce tip cooking for the art project illustrate contemporary Arctic art well, which Jokela et al. [9] define as New Genre Arctic Art. The practice revitalizes ecoculture for contemporary society and communicates the essence of traditional knowledge and environmental concerns. It also decolonizes the view of art—not to be seen as an object in a white

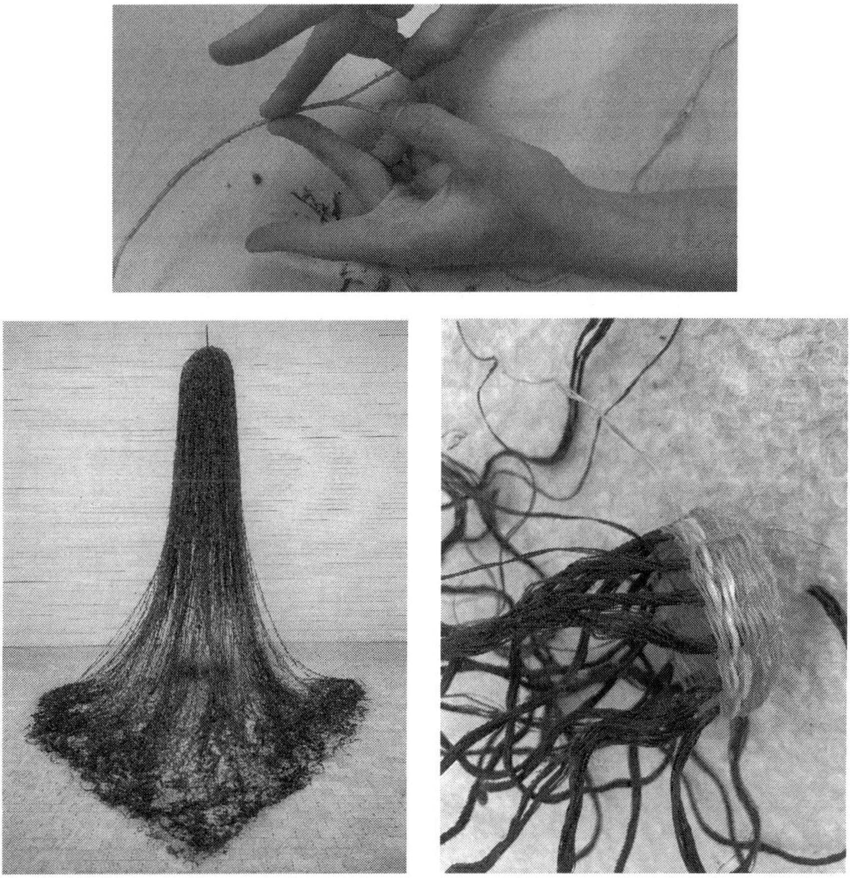

Figure 1.2 (*Left* side) Birgitta Linhart, "Skalp" (scalp) 2024. Photograph: Bjørnar Meisler, 2024; (*top* of figure) Photograph from the workshop by Birgitta Linhart, 2023; (*right* side) Photograph of weaving the polypropylene strings by Lande, 2024.

cube, but as part of the Northern lifestyle, rooted in closeness to nature and in practices of sharing within the human web of a natural ecosystem. Extensive root binding and spruce tip cooking culture could also be considered as decolonizing nature when used to stop invasive spruce from taking over the indigenous birch forest.

1.4.3 *Wool Embroidery for Science Communication*

Huhmarniemi revitalizes the tradition of blanket embroidery by dyeing wool yarn and stitching contemporary and political subjects on black woolen fabrics. She is a founding member of the Embroidered Stances Collective, committed to revitalizing blanket embroidery into expressive activism and craftivism [16].

Traditionally, black wool fabric and naturally dyed yarn were employed in the Nordic countries, with various decorative stitches defining and filling intricate patterns, often representing flora and fauna. The yarn's hues were derived from plant and mushroom dyes in a process that involved local and traditional knowledge of nature. The Embroidered Stances Collective continues the tradition of natural dyeing and thus also traditional knowledge for contemporary expressions and discussion. The collective presents embroideries as exhibitions and artistic research, thus decolonizing art from the Western separation of visual art and craft.

For the OOC project, Huhmarniemi embroidered a figure of the red rove beetle (*Olisthaerus substriatus*), which is on the so-called red list of endangered species. This species was reregistered for Junkerdalsura after restoration started [8].

Biodiversity research in spruce plantation removal aims to determine whether dead spruce wood from veteranization can be a resource for saproxylic (dead wood-dependent) beetles. The biological fieldwork was done in 2022 and will be followed up in 2025. Researchers consider whether veteranization could provide dead wood over a longer period and increase variation in dead wood that could be valuable for species associated with dead wood. Currently, more than 50% of the species in Norway live in forests and approximately 50% of the threatened species live in forests. Many of the threatened species in forests are insects (especially beetles) associated with dead wood.

Entirely different species of beetles were found in the birch forest. For these species to colonize the former spruce plantation, dead wood from deciduous trees must first be present. Dead wood is crucial in forest ecosystems for carbon storage, nutrient cycling, and supporting biodiversity. It serves as a key substrate for wood-dwelling beetles, aiding in wood decomposition and nutrient cycling. However, intensive forestry has reduced dead wood in production forests, negatively impacting these species.

The embroidery by Huhmarniemi celebrates the new registration of the red rove beetle. It follows the traditional style of presenting flora and fauna, but highlights insects and rotting wood in addition to typical themes of blossoming flowers. Embroidery acts as scientific communication and responds to the aim of strengthening the voice of New Genre Arctic Art in Arctic art-science collaborations.

1.4.4 School Participation

The second residency of OOC in the Junkerdal National Park included artistic work and workshops for school students to reflect on issues connected to biodiversity, nature conservation, and environmental ethics. The workshops were directed by Huhmarniemi and Linhart in the arboretum next to the Adde Zetterquist Art Gallery. By first introducing the students (aged 14) to looking at a problem from an artistic point of view, and through works of art addressing the environmental issues using art as a language, the students were split into groups and given the task of making a piece of art related to themselves and their place in a more-than-human nature.

As a key component of the project, the workshops increased dialogue with community members and left a lasting impact on the participating school students—they were energetic and eager to experiment with building and designing from natural materials. Presenting the artworks to everyone gave the students new food for thought, realizing that each group had addressed the given problem differently from the others. Participatory elements in the OOC aim to increase the impact of the project. A longer-term project with school communities will be carried out by a team of professors from Nord University and will be discussed in the following research about OOC. The participation of children as artists extends the acts of decolonization of art.

In a national park, human intervention in nature must be minimal. Michelkevicius [37] describes the Curonian Spit in Lithuania, a land formation managed as a national park and listed as a World Heritage site. He uses the term "green cube" to describe this type of cultural institution, where there is minimal human intervention. The green cube concept extends the idea of the white cube [38], describing how the landscape becomes timeless and detached from daily life, like an object in a museum. While the white cube isolates art from its context, art in the green cube is placed within its context and can even reach a point of total incorporation [37]. This view of art was introduced to the school students and extended to the aim of not harming nature.

The Norwegian philosopher Arne Næss developed the environmental-philosophical perspective of ecosophy, emphasizing the interconnectedness of all things and humans' ability to live harmoniously with the continuously growing and evolving environment [39]. By linking ecosophy, art in the green cube, and ecoculture, the term "green scape" emerges, referring to the integration of art within more-than-human nature, highlighting minimal human intervention, a deep connection to the more-than-human nature, and a reflection of colonization and decolonization. In the OOC, photography, installation art, crafting, and ecocultural revitalization were applied to document, respond to, witness, and reflect on environmental change and to consider and communicate what decolonization of art and nature means.

The art was carried out as New Genre Arctic Art that decolonizes art from Western categories and views. Decolonization of art means, in this case, artistic practice interwoven into ecoculture, craft-based methods, and lowering the hierarchy of who can be an artist. The following exhibitions by the OOC project will allow visitors to consider restoration and the human place in a more-than-human nature and to engage in discussions on biodiversity, human desires for nature, and the decolonization of nature and art.

1.5 Conclusion

Ecological conservation is complex and causes emotions when spruce trees are cut down and harmed. In this chapter, we demonstrated how art can help articulate the reasons for and values of nature's decolonization. Art is also used to communicate biodiversity research and revitalize ecoculture. Artists document, demonstrate, reflect, and comment on changes in nature in the Junkerdal

National Park area and tackle the complex questions of nature restoration in dialogue with national park managers and scientists to support the decolonization of Arctic nature and art. Art enhances the comprehension of ecological restoration as a process of decolonizing nature: removing the planted and farmed spruces and, therefore, protecting native plants. To further clarify decolonization efforts, the concept of New Genre Arctic Art is used to describe political and pedagogical art in the Arctic, and the term "green scape" is introduced to intertwine ecosophy, ecoculture, art, and national parks. The human role is seen as a partner in nature's biodiversity, actively supporting it rather than just avoiding harm through forestry and other extractive cultures. In this chapter, decolonizing art refers to the rejection of Western categories and views of art and covers the revitalization of Arctic ecoculture as part of curatorial and artistic production.

Acknowledgments

The OOC's institutional partners are the Adde Zetterquist Art Gallery, the Nordland National Park Centre, Midtre Nordland nasjonalparkstyre, the Norwegian Institute of Nature Research (NINA), the Norwegian University of Life Sciences (NMBU), the University of Lapland, Nord University, and the Arctic Sustainable Arts and Design Network. The Nordic Culture Fund funds research.

References

1. Higgs, E. The two-culture problem: Ecological restoration and the integration of knowledge. *Restoration Ecology*, 2005;13(1):159–164. [Accessed 29 June 2024]. Available from: https://doi.org/10.1111/j.1526-100X.2005.00020.x
2. Clark, S. E., Magrane, E., Baumgartner, T., Bennett, S. E. K., Bogan, M., Edwards, T., … Wilder, B. T. 6&6: A transdisciplinary approach to art–science collaboration. *Bioscience*, 2020;70(9):821–829. [Accessed 29 June 2024]. Available from: https://doi.org/10.1093/biosci/biaa076
3. Curtis, D. J. Creating inspiration: The role of the arts in creating empathy for ecological restoration. *Ecological Management & Restoration*, 2009;10(3):174–184. [Accessed 29 June 2024]. Available from: https://doi.org/10.1111/j.1442-8903.2009.00487.x
4. Harrower, J., Parker, J., & Merson, M. Species loss: Exploring opportunities with art–science. *Integrative and Comparative Biology*, 2018;58(1):103–112. [Accessed 29 June 2024]. Available from: https://doi.org/10.1093/icb/icy016
5. Huhmarniemi, M.. Art-based events for conflicted communities: Engaging and educating through art. *International Journal on Education Through Art*, 2021;17(2):271–280. [Accessed 29 June 2024]. Available from: https://doi.org/10.1386/eta_00065_3
6. Fritheim, P., Jacobsen, R. M. & Olsen, S. L. Fra granplantefelt til stedegen fjellbjørkeskog i Junkerdalsura naturreservat: Korttidseffekter av restaurering på vegetasjon [From spruce plantation to native mountain birch forest in the Junkerdalsura Nature Reserve: Short-term effects of restoration on vegetation]. *Blyttia*, (2024); 82(2):125–135.
7. Fritheim, P. Fra granplantefelt til stedegen fjellbjørkeskog i Junkerdalsura naturreservat: Korttidseffekter av restaurering på vegetasjon [From Spruce Plantations

to Native Mountain Birch Forest in Junkerdalsura Nature Reserve: Short-term Effects of Restoration on Vegetation]. Master's thesis, Norwegian University of Life Sciences; 2023. [Accessed 29 June 2024]. Available from: https://hdl.handle.net/11250/3076817

8. Laubo, L. Skogrestaurering i Junkerdalsura naturreservat: effekter på vedlevende biller [Forest Restoration in Junkerdalsura Nature Reserve: Effects on Wood-living Beetles] Master's thesis, Norwegian University of Life Sciences; 2023. [Accessed 29 June 2024]. Available from: https://hdl.handle.net/11250/3079323
9. Jokela, T., Huhmarniemi, M., Beer, R. & Soloviova, A. Mapping new genre Arctic art. In: L. Heininen; H. Exner-Pirot & J. Barnes, editors. *Arctic Yearbook 2021: Defining and Mapping Sovereignties, Policies and Perceptions*. Akureyri, Iceland: Arctic Portal; 2021. [Accessed 29 June 2024]. Available from: https://arcticyearbook.com/arctic-yearbook/2021/2021-scholarly-papers/400-mapping-new-genre-arctic-art
10. Jokela, T. & Huhmarniemi, M. Stories transmitted through art for the revitalization and decolonization of the Arctic. In: R. Sørly, T. Ghaye & B. Kårtveit, editors. *Stories of Change and Sustainability in the Arctic Regions: The Interdependence of Local and Global*. Abingdon, Oxon; New York, NY: Routledge; 2021. pp. 57–71.
11. Huhmarniemi, M. Marjamatkoilla ja kotipalkisilla. Keskustelua Lapin ympäristökonflikteista nykytaiteen keinoin [Artists in the Landscape of Berry Wars and Reindeer Husbandry: Contemporary Art as a Forum for Environmental Politics]. Doctoral thesis, University of Lapland; 2016. [Accessed 29 June 2024]. Available from: https://urn.fi/URN:ISBN:978-952-484-898-5
12. Raatikainen, K., Juhola, K., Huhmarniemi, M., & Peña-Lagos, H. "Face the cow": Reconnecting to nature and increasing capacities for pro-environmental agency. *Ecosystems and People*, 2020;*16*(1):273–289. [Accessed 29 June 2024]. Available from: https://doi.org/10.1080/26395916.2020.1817151
13. Jokela, T., Huhmarniemi, M. & Mäkitalo, T. Be the nature: Enhancing nature-connectedness through art museum pedagogy. In: A. Sinner, P. Osler, & W. Boyd, editors. *Propositions for Museum Education: International Art Educators in Conversation*. Bristol: Intellect; 2024. pp. 112–122.
14. Huhmarniemi, M. & Jokela, T. Arctic art and material culture: Northern knowledge and cultural resilience in the Northernmost Europe. In L. Heininen; H. Exner-Pirot & J. Barnes (Eds.), *Arctic Yearbook 2020: Climate Change and the Arctic: Global Origins, Regional Responsibilities?* Akureyri, Iceland: Arctic Portal; 2020. [Accessed 29 June 2024]. Available from: https://arcticyearbook.com/arctic-yearbook/2020/2020-scholarly-papers/350-arctic-art-and-material-culture-northern-knowledge-and-cultural-resilience-in-northernmost-europe
15. Huhmarniemi, M. Berry-picking culture and conflicts in Finland. *International Journal on Education Through Art*, 2024;20(1):51–67. [Accessed 29 June 2024]. Available from: https://doi.org/10.1386/eta_00151_1
16. Cervantes, F. & Huhmarniemi, M. Stitching for material sensitivity: From traditional to activist embroidery. *Ruukku – Studies in Artistic Research*, (2023): 20. [Accessed 29 June 2024]. Available from: https://doi.org/10.22501/ruu.1927154
17. Rousell, D., & Fell, F. 'Becoming a work of art' revisited: Ecologies of collaboration in tertiary visual arts education1. *International Journal of Education Through Art*, 2024;20(1):27–49. [Accessed 20 June 2024]. Available from: https://doi.org/10.1386/eta_00150_1
18. Leavy, P. *Method Meets Art: Arts-based Research Practice*. New York: Guilford Press; 2009.
19. Ballengée, B. *Ecological Understanding Through Transdisciplinary Art and Participatory Biology*. Plymouth: Plymouth University; 2015. [Accessed 29 June 2024]. Available from: https://doi.org/10.24382/4087
20. Polli, A. *Communicating Air: Alternative Pathways to Environmental Knowing Through the Experience of Geosonification and Other Ecomedia*. Plymouth: University of Plymouth; 2011.

21. Demos, T. *Decolonizing Nature: Contemporary Art and the Politics of Ecology*. London: Sternberg Press; 2016.
22. Chartier, D. *What is the Imagined North? Ethical Principles*. Québec, Canada: Presses de l'Université du Québec; 2018.
23. Gismondi, C. J. Why is the Arctic always white? Circumpolar Indigenous artists in the age of the anthropocene. *Esse*, 2019;97(97):92–95. [Accessed 29 June 2024]. Available from: https://esse.ca/en/young-critics/why-is-the-arctic-always-white-circumpolar-indigenous-artists-in-the-age-of-the-anthropocene/
24. Huhmarniemi, M. Making art in the land of polar bears and lemmings: Art and science expeditions in the Arctic. In: T. Hautala-Hirvioja, Y. Holt, & M. Mäkikalli, editors. *North as a Meaning in Design and Art*. Rovaniemi: Lapland University Press; 2019. pp. 182–194.
25. Decker, J. Climate of change. *Museum Management and Curatorship*, 2020;35(6):636–652. [Accessed 29 June 2024]. Available from: https://doi.org/10.1080/09647775.2020.1836999
26. Marsching, J. D., & Polli, A. (Eds.) (2012). *Far Field: Digital Culture, Climate Change, and the Poles*. Bristol, UK; Chicago, IL: Intellect; 2009.
27. Bloom, L. *Climate Change and the New Polar Aesthetics: Artists Reimagine the Arctic and Antarctic*. Durham: Duke University Press; 2022.
28. Härkönen, E., Huhmarniemi, M. & Jokela, T. Crafting sustainability. Handcraft in contemporary art and cultural sustainability in Lapland. *Sustainability*, 2018; *10*(6). [Accessed 29 June 2024]. Available from: https://doi.org/10.3390/su10061907
29. Jokela, T., Hiltunen, M., & Huhmarniemi, M. Art-based action research: Participatory art education research for the North. In: A. Sinner, R. Irwin, & J. Adams, editors. *Provoking the Field. PhDs in Education: International Perspectives on Visual Arts*. Bristol, UK; Chicago, IL: Intellect; 2019. pp. 45–56
30. McIntosh, P. *Action Research and Reflective Practice. Creative and Visual Methods to Facilitate Reflection and Learning*. London: Routledge; (2010).
31. Bradbury, H. *How to do Action Research for Transformations: At a Time of Eco-Social Crisis*. Cheltenham, United Kingdom: Edward Elgar Publishing; 2022. [Accessed 29 June 2024]. Available from: https://doi.org/10.4337/9781800373051
32. Isomursu, E. P. Explaining "real" in landscape photography. *SEA ART Education VISUAL Journal* IMAG #11, 2012;2021(11):10–16. [Accessed 29 June 2024]. Available from: https://www.insea.org/wp-content/uploads/2021/08/IMAG_11_ISOMURSU.pdf
33. Isomursu, E. P., & Jokela, T. Truthfulness in nature landscape photography. In: G. Coutts, & T. Jokela, editors. *Relate North: Possible Futures*. InSEA Publications; 2023. pp. 74–97. [Accessed 29 June 2024]. Available from: https://www.insea.org/wp-content/uploads/2023/12/RelateNorth-PossibleFutures_WEB.pdf
34. Paul, F. *Spruce Root Basketry of the Alaska Tlingit* (2nd repr. ed.). Sitka, Alaska: Sheldon Jackson Museum; 1991.
35. samer.se (n.d). Rotslöjd [online]. Samiskt informationscentrum [Sámi Information Centre]. [Accessed 20 June 2024]. Available from: https://www.samer.se/1155
36. Churchill, D. *Spruce-Root Basketry*. Juneau, Alaska: Sealaska Heritage Institute; 2019. [Accessed 29 June 2024]. Available from: https://www.youtube.com/playlist?list=PLGoKk-JZWo1NsOkcyra1nio-2f4nVfZO8
37. Michelkevicius, V. Artistic research in the Green Cube. In: F. Martini, & V. Michelkevicius, editors. *Tourists Like Us: Critical Tourism and Contemporary Art Practices*. Vilnius, Lithuania: Sierre and Vilnius Academy of Arts Press; 2013.
38. O'Doherty, B. *Inside the White Cube: The Ideology of the Gallery Space*. San Francisco: The Lapis Press; 1976.
39. Næss, A. *Økologi, samfunn og livsstil* [*Ecology, Society, and Lifestyle*]. Oslo: Bokklubben Dagens bøker; 1999. [Accessed 29 June 2024]. Available from: https://www.nb.no/items/URN:NBN:no-nb_digibok_2008090301002?page=65

2 Community Photography as a Tool for Witnessing Environmental Change and Contamination in Labrador

Jessica Penney and Eldred Allen

2.1 Introduction

The Inuit community of Rigolet, Nunatsiavut, is on the front lines of environmental change in Northern Canada [1]. Residents have been observing changes to the land and water for many years and have been involved in climate change research and initiatives to mitigate and adapt to a rapidly shifting environment [2–4]. Rigolet is also anticipated to be the most affected by the controversial Muskrat Falls hydroelectric project (MFP). The MFP is a large development constructed by the now-defunct Nalcor Energy and is currently managed by Newfoundland and Labrador (NL) Hydro. Both Nalcor and NL Hydro are former and current crown corporations of the provincial Government of NL, meaning they were created to meet public needs, but are run similar to private corporations at arm's length from the government [5].

The MFP is located about 30 kilometers downstream from the mouth of the Churchill River, also known as the Grand or Hamilton River, which flows into Lake Melville, a freshwater estuary which empties into the North Atlantic Ocean. The community of Rigolet is located on Lake Melville and relies on the estuary for hunting seals, fishing, and gathering other important food sources like seabird eggs. These harvesting activities also have important social, cultural, and health benefits to Inuit, as Inuit wellbeing is tightly interwoven with the health of the land and water, as harvesting activities reinforce a connection to culture, identity, and a feeling of self-reliance that is beneficial to physical and mental health [6].

The MFP has gained infamy in NL for its economic impact on the province, and the lack of transparency and accountability around its development. It was the subject of a public inquiry completed in 2020, focused on the economic and governance issues [7]. However, Inuit in Labrador have been outspoken about the potential environmental and health impacts of the Muskrat Falls project for several years, especially since research showed it could increase levels of methylmercury above recommended levels in important Inuit food sources such as seals, fish, and seabird eggs [8]. Methylmercury is a naturally occurring environmental contaminant that can be increased beyond safe levels with the construction of hydroelectric reservoirs as large areas of land are flooded [9]. It is a neurotoxin which bioaccumulates in animals and people and can have health

DOI: 10.4324/9781003545880-4

impacts such as cardiovascular and immune system issues, as well as impact brain development [10]. This is concerning for Inuit, as traditional foods not only have significance culturally but are also an important source of crucial nutrients like vitamin E, vitamin D, iron, zinc, magnesium, among others [11]. Despite significant resistance efforts from Indigenous peoples, including many Inuit from Rigolet, who asked for changes to the project's construction, the MFP was built without recommended mitigation measures. The potential physical health impacts related to the project are yet to be seen.

This chapter emerged from a partnership between the authors in 2020–2021, during Jessica Penney's doctoral research on the social and wellbeing impacts of the MFP. Jessica and Eldred Allen partnered on the creation of a participatory photography project that took place in person and online during the COVID-19 pandemic. Both authors are Nunatsiavut Inuuk, and Jessica is a researcher raised in Iqaluit, Nunavut, whose work focuses on the lived experience of environmental change in Inuit communities, specifically on Inuit food security and food sovereignty. Eldred is an entrepreneur and artistic photographer from Rigolet, where he owns and operates a drone/photography data collection business. His business is in demand from local and national clients, and he is represented artistically by renowned art galleries with photographs in many art collections.

There has been extensive scientific research on the biological and ecological impacts of the Muskrat Falls project [8], but little social science research on the lived experience of the Muskrat Falls project, especially from an Inuit-specific approach. Therefore, this research project aimed to understand how Nunatsiavut Inuit perceive the MFP and its potential impacts. Objectives were to explore the hydro project's impacts on Inuit conceptions of health and wellbeing, to understand desired health risk communication strategies for information about the MFP and to explore the relationship between the MFP and structures of colonialism in Canada. Photography was one of three methods used, in addition to semi-structured interviews and anonymous surveys (the latter two methods are not the focus of this chapter).

2.2 Photography as a Method for Discussing Community Concerns

This participatory photography project focused on themes that came from the community throughout the wider research project, in particular the social, cultural, health, and environmental impacts of the MFP. Participatory photography is an arts-based research method often used in community-based and participatory action research. One of the most popular methods is photovoice, which aims to bring participants into the whole research process, as participants take photos and discuss them together and with researchers [1, 12]. Photovoice has been used in other research projects related to health and Indigenous wellbeing [12, 13] and works from the basis that the images and stories generated during the photovoice process can be used as tools to engage policy and decision-makers [1].

This project was an adaptation of the photovoice method. Before the start of the project, all necessary equipment was shipped to Rigolet. Then, Eldred hosted two in-person workshops with six participants in Rigolet. The first workshop focused on how to operate the cameras. He taught the technical aspects of camera operation which included controlling exposure settings. Once a basic understanding of technical camera operation is understood, he moved on to the second portion which focused on photography techniques and storytelling through imagery. By manipulating exposure settings, such as shutter speed or aperture, the participants could express their feelings better in their photographs, making subjects blurred or out of focus. The remainder of the project took place online through social media. Prompts were posted in a Facebook group, where participants posted their photos and descriptions of the photos. In total, 26 photos and descriptions were submitted to the project group, including those by the authors, who also participated in the project. The remainder of this chapter draws on a number of these photos and some of the recurring themes which emerged.

The prompts included questions on the topic of environmental impacts, land-based activities, cultural and identity impacts, intergenerational relationships, and how the Muskrat Falls project fits into history in Labrador. Questions were broad in scope to allow for individual storytelling and interpretation. For example, one introductory question asked: "How do you feel about the Muskrat Falls project?" Another asked, "What aspects of cultural identity matter to you? What cultural impacts do you think the Muskrat Falls project may have on Inuit/Labradorians? What do you think about this?" On the topic of intergenerational relationships, the prompt asked, "What relationships are important to you? Does the Muskrat Falls project potentially impact the way you interact with others? With children, youth, adults, or Elders? How do you feel about this?"

The photographic contributions were analyzed holistically, using inductive narrative analysis to allow for the inclusion of storytelling, relationality, and reflexivity [14]. In narrative analysis, the data (in this case photos and text) is coded in large pieces to retain the participant's story or narrative. In this way, the participant's story retains its intentional context. Narrative analysis is particularly attuned to social change and social justice with the relation between the narrator and audience as central to the research process [14]. This method of analysis recognizes the benefits of storytelling for the speaker, of the power of being heard, and the ways that collective stories have the ability to demand change, and how stories can create public dialogue [14].

2.3 Photography as Environmental Monitoring

In this project, participants' photos served as a way of witnessing environmental change and even monitoring recent shifts in local conditions. For example, the following participants' photos reference ice quality and formation. This is a concern, as many people fear the water and ice quality will change if salinity is affected by the managed flow of freshwater through the hydroelectric dam.

Ice formation is of the utmost importance in Inuit communities, as it is used for travel routes, and unexpected changes to water conditions may thin or weaken the ice and put people at risk of falling into the water.

In response to the introductory question posed to all participants during the photography project: "How do you feel about the Muskrat Falls project?" Rochelle Rich shared the photo in Figure 2.1 and wrote:

> I feel that the impacts are far too negative and the claims that it's non-invasive are a huge lie. See the color in this shore ice? It's wrong, and all along the shoreline [where] it is seen the sea ice that usually forms EVERY Winter is not able to form. There is too much sediment there for the snow, sea water and wind to make good ice.

Another participant, Paula McLean, shared the photo in Figure 2.1 and the following description:

> The pictures I chose was taken the morning of February 2, 2021.
>
> Seeing the vapor on the water, representing a colder temperature than what we have been experiencing gave me a feeling of happiness and hope that the colder temperatures have finally arrived. We have been having such a mild winter thus far, with not a lot of good solid ice forming for our Winter highway or Spring travel. The things that are vital to our community, for many reasons such as harvesting wildlife, gathering wood, good mental health and peace of mind, to name a few.

Figure 2.1 *Top left*: Ice along the shore. Photo by Rochelle Rich. *Top right*: Vapor on the water. Photo by Paula McLean. *Bottom left*: Newly formed ice. Photo by Katheline Pottle. *Bottom right*: Camouflaged hare behind a tree. Photo by Katheline Pottle.

> So needless to say watching the vapor rise in the morning was a great way to start the day!

A third participant, Katheline Pottle, submitted the photo in Figure 2.1 and description:

> The picture I chose is newly formed ice. To me this indicates how the "outside world" is [affecting] our small corner of the world. It depicts the changes of increased salinity in our water/ice affecting our culture and traditions. This I believe is a direct result of Muskrat Falls, as well as the effect of global warming. In my experience February 1 would typically be colder, the ice would be thick and covered with snow. Today I took pictures with bare hands, not something one would normally do in the winter time. This photo is taken "up along the shore." We can only see clearly directly in front of us; things in the distance are unclear. We have an idea of what is there; however, the future is blurry. Hopefully we will have a hand in focusing our future to the benefit of our people and beautiful land.

The photos and descriptions above function as a form of environmental monitoring. The photos show changes to ice and water conditions, and descriptions contextualize the observable changes to the environment that participants have witnessed firsthand during their lifetime. The photos also bring together issues of climate change and the impacts of the Muskrat Falls project, which are happening simultaneously and should be considered in tandem.

Environmental monitoring in the North is often performed through methods such as animal and water sampling and is frequently removed from community perspectives and Indigenous Knowledge, though there is a rapidly growing effort—and acknowledgement of the necessity—to include Indigenous Knowledge in monitoring activities [15, 16]. However, the participants in this project actively monitor their community, land, and water visually through their photos and personal observations, showing how arts-based methods can be another lens through which to monitor environmental change.

2.4 Photography as a Method of Witnessing Change

Photography also functions as a way of documenting change and loss, both personally and at a wider scope. It preserves the present, addressing feelings of powerlessness, and documents the social, cultural, and environmental challenges at hand due to circumstances outside the community's control.

As Katheline Pottle wrote alongside the photo of the camouflaged hare behind a tree in Figure 2.1:

> My family and community are important to me.
>
> Since the Muskrat Falls project I now hear people comment more frequently (almost wondering out loud), "I guess it's not okay to eat the fish …

> probably for not much longer now." Conservation seems to have taken on a lesser meaning – due to the fact that safely consuming fish and other foods is now an unknown. I feel like we are running on a treadmill and the speed keeps increasing. Running faster and faster to keep up and unsure how much longer we have before we are thrown off – ending up God knows where and in what condition.

One of Eldred Allen's contributions, seen in Figure 2.2, discussed the future safety of fish, an important food source:

> As an image to represent the first question I captured this image on January 30. One of the main concerns of Rigolet residents is how Muskrat Falls will have a detrimental effect on our fish stocks and the safety to consume them. This image of a trout nearly buried in snow and shrouded by the drifting snow represents how our fish are fading and are becoming obscure. We can still make out what they are, feel great excitement to see them but how long will they be there before they are completely gone?

Another participant, Rhoda Palliser, discussed thinking about the future for her family and future generations in a photo taken from her living room window (see Figure 2.2):

> The first picture I [chose] was a very clear picture of the view from my living room window. One I think can speak for what once was. How we

Figure 2.2 *Top left*: Trout buried in snow. Photo by Eldred Allen. *Top right*: View from living room window. Photo by Rhoda Palliser. *Bottom left*: Double exposure of a corporate machine turning a blind eye. Photo by Eldred Allen. *Bottom right*: Fly on the window. Photo by Katheline Pottle.

> fished and hunted out in that water for many many years, with no care that maybe one day we won't be able to or if we do there might be a chance that the fish we catch or the seal we shot can be poisonous to us. Which really makes me sad. Sad for myself, my husband, my children and later on the ones who would be affected the most my future grandchildren.

These photographs and texts highlight how, for Inuit, personal change is intimately connected to environmental change. According to participant photographers, individuals, families, and the community will transform alongside altering ice, waterways, and food sources.

This knowledge shared by participants aligns with existing research on the ways that environmental health is a determinant of, and a protective factor for, Inuit health. Power writes that "Food obtained from traditional food systems links the environment and human health, and forms the basis of social activity, social cohesion, and social integration" (p. 96, [17]). This is not only true historically but also, as the participant photographers allude to, Inuit foods continue to be important. Research studying harvesting activities related to several types of seabirds, salmon, Arctic char and caribou found that "approximately 85% of households are active harvesters in at least one animal from those species studied" (p. 166, [18]). These harvesting practices provide outstanding nutritional benefit, as research by Calder et al. found that, for Inuit in Rigolet, traditional foods made up 4% of total calories, but up to 70% of nutrients consumed [19]. Thus, the photos and descriptions highlight the real health risks of a potential shift in diet for Rigolet Inuit, as well as the resulting associated cultural and emotional loss to future generations.

2.5 Photography as a Form of Agency

Participatory photography can also be a way of reclaiming power and agency—becoming an outlet where one can share their perspective despite being constrained by structural issues. Plenty has been written about photojournalists and photography witnessing political violence, war, and other forms of conflict through their camera lens [20–22]. In this case, photography allows the participant-photographers to tell stories that are important to their everyday lives. Further, the photos and descriptions in this project serve as a method through which to reflect upon historical oppression and witness the inequities their community and region face today. Another of Eldred Allen's photos in Figure 2.2 highlights these structural issues associated with corporate exploitation of the environment:

> Rigolet residents lead the protests in Goose Bay as we have so much to lose. The image below sums up the Muskrat Falls project and how a lot of us feel towards it. This image is a double exposure. It illustrates the

> next generation who have so much to lose, they were involved in the protests and they stand there looking at the waters which will be polluted. All the while the Corporate machine is transparent, not really present during all of this; they have their back turned to what protesters are looking to protect and they are turning a blind eye to the residents and their concerns.

Katheline Pottle also highlighted how the MFP was one example in a long line of policy decisions in which the Government of NL has not prioritized Labradorians.

Alongside a photo in Figure 2.2, she wrote:

> I feel that the Muskrat Falls project is another example of the dismissive attitude that our provincial government has towards our community, our people and Labrador in general.
>
> We are like the fly on the window, the focus is everywhere else. We are nearly invisible.

The GNL has a long history of supporting extractive industries in Labrador, including mining, forestry, and hydroelectricity. Some people feel that these industries have largely benefited the island of Newfoundland, rather than the people of Labrador.

Despite these challenges and tensions between the governments in Inuit, the photo below by Lorraine Allen shows how communities come together to defend their rights and interests, and the description highlights hope for how this can happen again in the future. In particular, the image references an intense period of protest against the MFP that happened in October 2016, and how that might be revived if another planned hydroelectric project near Gull Island is constructed (Figure 2.3).

> Power To The People
>
> The development of the Muskrat Falls Project was shrouded in controversy from the beginning. Labradorians from all walks of life, young and old, were a part of the movement – Make Muskrat Right, an attempt to halt this project. Protests were held in Rigolet, Goose Bay and at the Nalcor gates (the work camp for the project). The gates were stormed and a takeover ensued, becoming the focal point of this historic event. This movement forged positive relationships between the Indigenous groups and other Labradorians alike. This one cause fused all of Labrador. It was the ultimate in relationships and continues to serve as a reminder to what 'the people' can accomplish when taking a stand together. Now with talks of the development of Gull Island, I for one would be incensed and proud to once again take up the cause, alongside fellow Labradorians. To do my part to try to put a stop to another mega hydro project from ruining this land God gave to Cain.

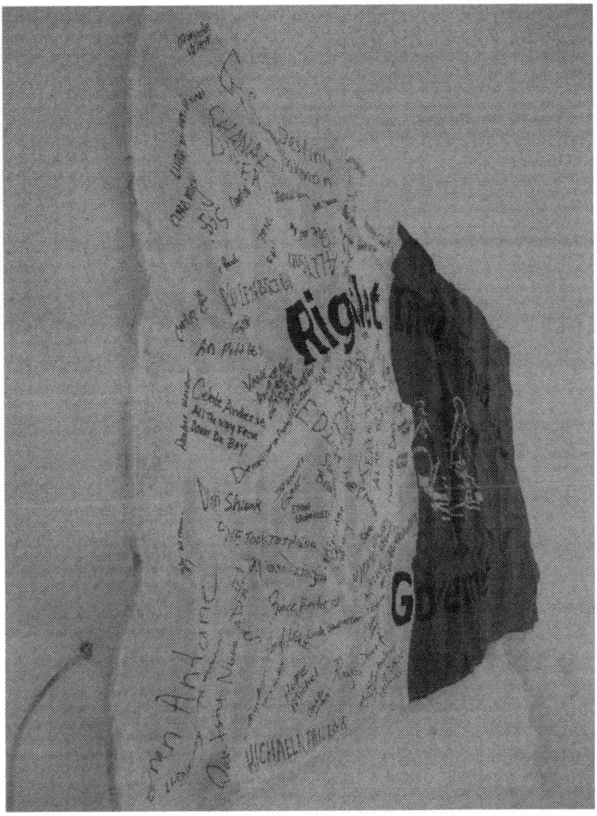

Figure 2.3 Rigolet Inuit Community Government flag. Photo by Lorraine Allen.

The above photo and description show how, in addition to highlighting social and environmental issues, participatory photography can put forward aspirations. It is a way to highlight collective efforts to a wide audience and be a positive tool for advancing community desires in the face of adversity.

2.6 Conclusion

Overall, the photos from the Rigolet participatory photography project clearly illustrate how photography can be a powerful visual tool for personal expression and agency in the context of current and future threats to the land and water. Photography allows the photographer to use their agency to tell stories important to them. The resulting photos powerfully convey the emotional toll and lived experience of environmental change emerging from energy development in and around Indigenous communities. The photos show both drastic and subtle changes and highlight concerns and fears, while also representing hope for the future.

References

1. Palibroda, B, Krieg B, Murdock L, Havelock J. *A practical guide to photovoice: Sharing pictures, telling stories and changing communities*. Winnipeg (MB): Prairie Women's Health Network.
2. Willox AC, Harper SL, Edge VL, Landman K, Houle K, Ford JD. The land enriches the soul: On climatic and environmental change, affect, and emotional health and well-being in Rigolet, Nunatsiavut, Canada. *Emotion, Space and Society*. 2013;6:14–24.
3. Harper SL, Edge VL, Cunsolo Willox A, Rigolet Inuit Community Government. 'Changing climate, changing health, changing stories' profile: Using an EcoHealth approach to explore impacts of climate change on Inuit health. *EcoHealth*. 2012;9:89–101.
4. MacDonald JP, Willox AC, Ford JD, Shiwak I, Wood M, Government RI, IMHACC Team. Protective factors for mental health and well-being in a changing climate: Perspectives from Inuit youth in Nunatsiavut, Labrador. *Social Science & Medicine*. 2015;141:133–141.
5. Bernier L, Dutil P, Hafsi T. Policy adrift: Canadian Crown corporations in the 21st century. *Annals of Public and Cooperative Economics*. 2018;89(3):459–474.
6. Inuit Tapiriit Kanatami. Social determinants of Inuit health in Canada. Inuit Tapiriit Kanatami; 2015.
7. LeBlanc RD. Muskrat Falls: A misguided project. Commission of Inquiry Respecting the Muskrat Falls Project; 2020.
8. Calder RS, Schartup AT, Li M, Valberg AP, Balcom PH, Sunderland EM. Future impacts of hydroelectric power development on methylmercury exposures of Canadian indigenous communities. *Environmental Science & Technology*. 2016; 50(23):13115–13122.
9. Mailman M, Stepnuk L, Cicek N, Bodaly RD. Strategies to lower methyl mercury concentrations in hydroelectric reservoirs and lakes: A review. *Science of the Total Environment*. 2006;368(1):224–235.
10. Schartup AT, Balcom PH, Soerensen AL, Gosnell KJ, Calder RS, Mason RP, Sunderland EM. Freshwater discharges drive high levels of methylmercury in Arctic marine biota. *Proceedings of the National Academy of Sciences*. 2015;112(38): 11789–11794.
11. Kuhnlein HV, Receveur O. Local cultural animal food contributes high levels of nutrients for Arctic Canadian Indigenous Adults and Children1. *The Journal of Nutrition*. 2007;137(4):1110–1114.
12. Poudrier J, Mac-Lean RT. 'We've fallen into the cracks': Aboriginal women's experiences with breast cancer through photovoice. *Nursing Inquiry*. 2009;16(4): 306–317.
13. Nykiforuk CI, Vallianatos H, Nieuwendyk LM. Photovoice as a method for revealing community perceptions of the built and social environment. *International Journal of Qualitative Methods*. 2011;10(2):103–124.
14. Chase, S. E. 2011. Narrative inquiry. In: Denzin, N. K. & Lincoln, Y. S. (eds.). *The Sage Handbook of Qualitative Research*. 4th ed. Thousand Oaks: SAGE Publications, Inc. 421–434.
15. Thompson KL, Lantz TC, Ban NC. A review of Indigenous knowledge and participation in environmental monitoring. *Ecology & Society*. 2020;25(2):10.
16. Wheeler H, Danielsen F, Fidel M, Hausner VH, Horstkotte T, Johnson N, Lee O, Mukherjee N, Amos A, Ashtorn H, Ballari Ø. The need for transformative changes in the use of Indigenous knowledge along with science for environmental decision-making in the Arctic. *People and Nature*. 2020;2, 544.
17. Power EM. Conceptualizing food security for Aboriginal people in Canada. *Canadian Journal of Public Health*. 2008;99(2):95–97.

18. Felt L, Natcher DC, Procter A, Sillit N, Winters K, Gear T, Winters D, Nochasak S, Andersen S, Ford R, Flowers H. The more things change: Patterns of country food harvesting by the Labrador Inuit on the North Labrador Coast. *Settlement, Subsistence, and Change Among the Labrador Inuit: The Nunatsiavummiut Experience*. 2012;2:139.
19. Calder RS, Bromage S, Sunderland EM. Risk tradeoffs associated with traditional food advisories for Labrador Inuit. *Environmental Research*. 2019;168:496–506.
20. Möller F. Witnessing violence through photography. *Global Discourse*. 2017;7(2–3): 264–281.
21. Möller F, Sontag S. Rwanda revisualized: Genocide, photography, and the era of the witness. *Alternatives*. 2010;35(2):113–136.
22. Duffy A. Bearing witness to atrocity crimes: Photography and international law. *Human Rights Quarterly*. 2018;40(4):776–814.

3 Sensing, Storytelling, and the Sacred

Two Creative Multimedia Projects Exploring Arctic Change

Chris Dunn

3.1 Introduction

I have visited, worked, and traveled widely in the Arctic on and off for over a decade—in Iceland, Alaska, and a summer in Greenland. Nevertheless, my perspective on this vast region remains limited—temporally, spatially, and culturally. Through the windows I have been fortunate to peer through, I have sought to creatively, yet authentically, capture the changes occurring in the Arctic. In the following chapter, I describe two disparate, yet related projects, either centered around or incorporating the unfolding realities of climate change. I discuss my artistic process, the underlying motivations and philosophical foundations for each project, and my struggle as a relative outsider, with a limited perspective, to document Arctic change.

I include the term sacred in the title of this chapter for two key reasons. First, it is a term frequently used by Indigenous peoples to describe places and elements of their homelands [1–5]. One of the seminal texts on traditional ecological knowledge, *Sacred Ecology*, draws a direct connection between the term sacred and the infusion of a "life-force into the machine-like scientific conceptualizations of ecosystems" [6, p. 12]. In this sense, the sacred stands in contrast to reductionist ecological frameworks, instead suggesting an integrated relationship between humans and the environment—one rooted in care, reciprocity, and relationality.

A sense of the sacred may also arise from an intuitive, affective response to particular places and things, an orientation shaped by gratitude and reverence. As one author puts it, "reality engages us rightfully as the sacred" [7, p. 194]. This orientation aligns with what some have called a moral topography or moral cosmology [8]— a meaningful and reverent relationship to a place, deeply engaged with interconnected materiality, as opposed to a detached, universal stance, common in science, technology, and economics [9].

Second, "sacred" is a concept that I have often sought to investigate through academic and artistic projects. A key element of this inquiry is the relationship of sacredness and materiality, part of my broader interest in the human relationship with materiality—the constituent components of our world that we regularly interact with, that we shape, and that in turn shape us. This interest

DOI: 10.4324/9781003545880-5

encapsulates both the nonhuman world—nature, in a word—and our own artifacts and artificial environments. An element of this is how we conceptualize the environments we inhabit and visit, and our place in them. A fundamental question is whether the natural world is merely a storehouse of resources or something vital to our lives.

The sacred expresses this latter possibility in recognizing a sense of integrity and meaningfulness in things that are not a mere projection of human capriciousness or utilitarian consumptiveness. The Arctic context—where rapid change is occurring, where Indigenous peoples have long regarded elements and locales within their local landscapes as sacred, and where vast areas remain undeveloped, though with certain notable exceptions—stands out as a region in which to explore these issues.

3.1.1 Sensing Ice: Explorations of Knowing Nature

For this project, I spent a summer in Greenland and Iceland documenting glaciers and ice sheets through photography, supplemented by audio, video, and writing. While melting glaciers became a central component of this journey, the intended primary raison d'être for the exhibit was a philosophical exploration of how nature is known and represented, specifically contrasting "direct" or unmediated (especially technologically) embodied and sensorial encounter with ice, with the abstract and highly mediated portrayals of modern science through remote sensing, and distillations into graphs and numbers. This resulted in an immersive multimedia journey for exhibit visitors, produced in collaboration with scientists, who supplied visually striking, data-saturated remote imagery.

What unites this with project with *Raven's Light* is an exploration of my own sense of the sacredness of sensory encounter, how this sense might find points of intersection in other cultures, and how it is challenged by modernity, industrial development, and environmental change [10]. I have long sought to follow the examples of writers and thinkers like H.D. Thoreau and Barry Lopez in approaching the natural world—to peer deeply into things and places, giving them my attentiveness to allow them to meaningfully present themselves. Lopez sums this up powerfully in his magnus opus *Arctic Dreams*:

> The land (and ocean) retains an identity of its own, still deeper and more subtle than we can know. Our obligation toward it then becomes simple: to approach with an uncalculating mind, with an attitude of regard. To try to sense the range and variety of its expression—its weather and colors and animals. To intend from the beginning to preserve some of the mystery within it as a kind of wisdom to be experienced, not questioned. And to be alert for its openings, for that moment when something sacred reveals itself within the mundane, and you know the land knows you are there [11].

I have in turn tried, through my art and writing, to capture this process and its potency.

3.1.2 Raven's Light

I have visited many Alaskan Arctic villages over the course of several years, usually briefly. The most substantial period I spent in a single village was most of a summer in the inland Inuit community of Anaktuvuk Pass, investigating how subsistence access into the surrounding park landscape may be impeded, mainly due to shrinking snow cover caused by climate change. This gave me a window into this unique community, which has been responding to multiple dimensions of social and environmental change for over a century, at least since the age of exploration and whaling began to fundamentally alter this region. I produced a photographic collage based on the time I spent in Arctic Alaska. The collage explores the tensions between the traditional subsistence-based cultures and economies of the Indigenous peoples of Arctic Alaska and the globalized consumer culture that exists alongside these traditions, yet in certain respects, including through localized industrial impacts and climate change, threatens to undermine them. This exemplifies the conundrum we all find ourselves in, though Arctic peoples are uniquely vulnerable.

3.2 Theoretical Foundations

3.2.1 Sensing Ice: Explorations of Knowing Nature

What is it to know something—knowing, rather than simply knowing about? Arctic ice, in the form of glaciers, ice sheets, and sea ice, is, for most people, far-off and apart, something that occasionally appears in the media as a distant threat, though of course with global, and therefore local, implications.

The inquiry into what properly constitutes knowledge—epistemology—is a key subdiscipline of philosophy. One element of this conversation is the role of the senses in knowledge. Western thought and other traditions, for example, the Hindu concept of *maya*, have often considered the senses to be unreliable or even deceitful. One implication of this view is that mathematical reasoning and evidence are believed to be epistemologically superior to the senses—a view sedimented into modern science. Science, though empirical, is not typically *sensual* (nor experiential) as the ultimate standard of knowing is the tearing of data out of its felt context into the abstractions of graphs, equations, and models. Phenomenology is a philosophical tradition that prioritizes the senses as a deliberate counterbalance to modern science and its philosophical foundations [12].

Concerns over what constitutes genuine knowing also manifest, particularly in the Arctic context, in the form of considerations about knowledge held by Indigenous peoples, often termed traditional ecological knowledge or occasionally Indigenous science. This is a form of knowing that is primarily grounded in practical encounters with landscapes, rooted in specific places, and often accompanied by relational ontologies. It shares with phenomenology its profoundly sensual character—grounded in deep attentiveness to the world as a matter of survival and flourishing. Ice is a prominent feature of

Arctic landscapes and Indigenous lifeways, which has been considered in depth in anthropological works, notably *The Meaning of Ice: People and Sea Ice in Three Arctic Communities* [13]. Ice constitutes a hunting ground, building material, and travel corridor, as well as a site of ontological significance.

Though most of this project's media is from Greenland, which has a majority Indigenous population, I chose not to focus on these same themes, but rather on the purely sensual dimension of ice, exploring it as an aesthetic phenomenon—as a step toward developing a cryophenomenology. I also focused largely on landscapes of terrestrial ice, which are uninhabited and rarely visited. With logistical assistance from Greenlanders, among other partners, I produced an exhibit intended to be a philosophical exploration of how nature is known and represented, specifically contrasting first, "direct" or unmediated embodied and sensorial encounter with ice stemming from actually being in and attending to these icy locales, with second, the abstract and highly mediated portrayals of modern science through remote sensing, and distillations into graphs and numbers. What sort of gap, in other words, emerges between sensual reality and its scientific portrayal?

This was my first exploration of my academic concerns through art. Through this work, I sought to address questions I had long grappled with, including how devices and instruments—forms of technological mediation—shape our perception and in what sense might this mediation have moral significance. I also wondered: Is there a moral dimension to knowing? Are there moral underpinnings of sensual encounters? How does embodiment relate to knowing? To what extent is the disenchantment hypothesis—sociologist Max Weber's famous account of the modern process of rationalization and bureaucratization leading to the perception of a flattened world empty of resounding and resonant meaning—true, and, if so, what are its implications? That is, to what extent does the objectivizing dimension of science and technology produce a distancing and controlling relationship with nature? I approached these questions first by spending time immersed in Arctic locations with them in mind while seeking to deliberately attend to my experiences, and through the artistic practices of writing and photography in specific relation to my understanding of relevant scientific practices.

Sensing Ice was partly inspired by Barry Lopez's essay, *Learning to See* [14], in which he recounts giving up photography based on an encounter with a polar bear. Through this experience, he comes to explicitly recognize the mediating role of the camera—an obstruction to his attentive immersion in a wild encounter. I had read the essay years before and, like Lopez, decided to devote myself to wordcraft—conceptualizing it, perhaps mistakenly, as a less mediated form of expression and representation. Yet, over time, I picked up the camera despite myself—and its prominence has only grown in my work. Another inspiration was the film *Chasing Ice* [15], highlighting James Balog's glacial photography [16]. However, though ice is a prominent causality of climate change, and the exhibit featured dramatic scenes of melting ice and meltwater streams, my primary intention was not, like the film, to raise awareness of climate change.

3.2.2 Raven's Light

The native people of Arctic Alaska live in one of the Earth's more remote locations yet are centrally bound in complex global dynamics. The monetary economy of Alaska's North Slope is almost entirely based on fossil fuel extraction, while the effects of climate change caused by these very industries are felt disproportionately here: it is the fastest warming part of the Earth.

In this exhibit, raven is our guide. This is appropriate as raven is a central figure in many Indigenous cultures, including for the Inuit of Arctic Alaska. I can only claim to have a cursory knowledge of the significance of ravens in Inuit culture; this collage was thus the creative interpretation of an outsider. In Inuit lore, raven is a complex figure who is both creator and trickster, a shape-shifter who brought light to humans, played an important role in creating the landscape and social organization, and yet is often foolish and duped by other creatures.

One story tells of human destructiveness: people were killing all the animals, so raven hid the sun in a bag. Eventually, he was tricked, and the sun was restored. Another tells of raven's impatience: raven was turned from black to white when loon threw soot in response to raven's hasty tattooing. Others tell of raven's failed attempts at joining geese on migratory flights—drowning in the sea along the way. At a local level, however, raven has often been conceived as nearly omniscient—powerful in part because of its ability to fly and thus acquire knowledge inaccessible to humans.

Figure 3.1 Images from *Raven's Light* including collage displayed as part of *Mapping Home/Collecting Truths: Works by Indigenous and International Artists.*

This collage tells a new story where humans have attained and surpassed raven's power of flight, the enhanced perception this allows, as well as the power to bring light into darkness. The collage begins with ravens soaring through an empty blue sky, positioned beside an image of a rainbow. At the bottom, the arc of the rainbow is inverted, transformed into an oil pipeline that channels and contains the sun's energy. This visual suggests the latent human capacity for destruction can no longer be held in check by the raven's power over light.

Raven has been outshined by Ravn Alaska—otherwise known as Corvus Airlines—one of the major air carriers in the state, which serves remote communities like those pictured in the piece. These villages are off the road system in "bush" Alaska. The airplane is thus the primary connection to link distant locations and to move goods and people. It also means that things tend to arrive but never leave.

I worked in Anaktuvuk Pass on a small project focused on documenting local climate change observations, particularly at the intersection with subsistence—a direct dependence on the land for livelihood through hunting, fishing, and gathering. When I asked Elders about changes, they would often begin telling me about settling in the village, and later moving from sod into wooden houses, or transitioning from dog teams to snowmachines. I would ask specifically about changes to the landscape. Yet, as contemporary forces such as the airplane, imported food and goods, oil development, and others have proceeded in tandem with climate change, it is difficult to separate a specific change from the many it is connected to.

The village of Anaktuvuk Pass began in 1949 when the Nunamiut Iñupiat—the People of the Caribou—one of the last semi-nomadic peoples of North America, settled on a broad pass with ample willows and vast herds of migrating caribou. Anaktuvuk is an anglicized form of the Inupiaq place name, which means the place of caribou droppings. A key reason for settling was to gain access to the goods and services brought by airplane. The first great industrial incursion into Arctic Alaska was the Trans-Alaska Pipeline and support road connecting Prudhoe Bay on the Arctic Ocean to Valdez on Prince William Sound. Bulldozers roared into Anaktuvuk Pass in 1969, sent by Governor Wally Hickel. Ultimately, the pipeline, completed in 1977, followed a different route, thus maintaining the community's relative isolation.

Today, the large caribou migrations of the past have all but ceased coming through Anaktuvuk Pass. The reasons for this are complex, but almost certainly relate to a variety of industrial developments, including the Trans-Alaska Pipeline and supporting road, which enables easy access to the area by outsiders; noisy industrial development in Anaktuvuk Pass village; new hunting practices enabled by mechanized transportation like snowmobiles and ATVs; and climate change, which is driven by industrial development on a global scale, including the oil extraction taking place on the North Slope. In appropriating raven's power, we have traded away parts of raven's creation.

3.3 Methodology

3.3.1 Sensing Ice: Explorations of Knowing Nature

The cryosphere, the totality of the world's snow and ice, as a global phenomenon and a prominent casualty of climate change, offers a powerful proxy for our grasp of nature writ large. Glaciers, in their size, beauty, danger, and relative inaccessibility, seize the imagination, serving as inspiration for everything from Mary Shelley's *Frankenstein* to John Muir's "glacial gospel" to the "social ontology" of glaciers found in many Indigenous cultures.

Simultaneously, glaciers and ice are intense objects of study that are often measured remotely by satellites, translated into mass balance equations and projections of sea level rise, and funneled into technical papers and IPCC reports. Glaciers flow. Over bedrock, into oceans, and through various mediums of analysis and understanding. Glaciers bear down on our future, demanding our attention and our planning for their contingencies.

The Earth's ice is melting. Sea ice is diminishing while ancient ice caps and glaciers pour themselves into the oceans. To most of us, the Earth's icy landscapes seem remote. Yet the furthest reaches of the Earth are responding directly to us, to our daily decisions. Our fates are tightly bound together: as we shape the future of ice, ice shapes our future.

Figure 3.2 Images from *Sensing Ice: Explorations of Knowing Nature.*

Our daily lives now have a disproportionate and obvious impact on the furthest, most inaccessible reaches of the Earth. Glaciers and ice sheets in Greenland, Antarctica, and at the heights of the Himalayas unequivocally "respond to human behavior." Glaciers pour their bodies into the seas, transforming currents, creeping into coastal megalopolises, thereby imperiling our own bodies.

This exhibit was an immersive multimedia experience structured around a spatial journey of descent from the world's highest glaciers surrounding Mount Everest to the ocean by way of Greenland and Iceland, following the course of terrestrial ice as it melts and contributes to rising seas. Along the way, the exhibit allowed viewers to contemplate the evolving human relationship with ice and how this is informed by ways of knowing—both sensual and remote—the latter meaning distantly captured and technologically mediated.

In the showcase image, I am straddling an abyss with a camera held tight against my face. The camera is a constructed intermediary between me and the emptiness below. I stand above a water-filled crevasse photographing my reflection beneath. Ice in translucent surface blues transitions to black in the crevasse's impenetrable, narrowing depths. I look through my mirrorless digital camera's internal video monitor at my own reflection in nature's mirror—the surface of still water. I am looking for nature—immediate and unfiltered—and find a digital clone of my reflected self in black and blue staring back at me. In the resulting photograph, the dark abyss below runs longitudinally through my reflection, matching my body's curves, splitting me into two. I displayed this photograph directly alongside one of myself taking this same photo. In another photograph, a spearhead-shaped chunk of clear glacial ice melts in my hand against a blurred backdrop of the glacier from which it emerged. The image is symbolic of the contemporary human relationship with the cryosphere and nature more broadly.

The exhibit was multimedia, primarily photography, and intended to be multisensory. It included three video monitors in various locations, each showing video loops, a listening station with sounds of melting glaciers, and the option for visitors to listen with their own headphones to musical accompaniment by Alaskan composer Matthew Burtner incorporating glacial sounds [17].

The prints were deliberately kept free of titles and explanations, while most of the exhibit was without text. This was meant to emphasize the sensory—to allow ice to be engaged with as an aesthetic phenomenon rather than a causal mechanism or vehicle for analytic explanation. This reveals a fundamental paradox: the exhibit's exploration of mediation is insuperably limited by its own inherent mediation. Carefully selected light captured by the camera is first processed, then materialized on paper or as a digital display. This intention nevertheless shaped the exhibit's form.

I played with this separation with a few "hybrid" images where scientific data was overlaid on photographs, including a graph of Greenland's ice sheet mass balance over time with a supraglacial stream whose tilt paralleled that of the graph, and a map showing Greenland's surface melt with a rushing cascade of gray-brown sediment-filled meltwater. Certain of the images were visual representations of scientific data but were displayed to emphasize their aesthetic properties and were kept to the peripheries of the main display. A personal reflection on ice loss from a scientist at the National Snow and Ice Data Center and another from an Icelander at a University of Iceland satellite research center close to Iceland's largest ice cap emphasizes the "subjective" dimension of knowing or the relational quality of perceiving—of spending time directly engaged with ice.

Scientific explanations, identifications of ice features, and other details were included in a set of binders placed throughout the exhibit. Very few scientific descriptions were displayed on the walls. One notable exception was a diagram of glacial hydrology, identifying features and dynamics like moulins, crevasses, and basal flow. This information helped to emphasize the feeling of depth I sought to evoke, to pull visitors down and into the ice. The photos began with tiny cryoconite holes and expanded—deeper and wider—into moulins and crevasses; they also highlighted the varying forms and textures of ice.

The exhibit was displayed at a university library, which spans two floors connected by a staircase [18]. The feature of depth was the exhibit's core structure. The library's entrance was on the upper floor allowing visitors to embark on a journey of descent from the world's highest glaciers surrounding Mount Everest to the ocean by way of the Greenland ice sheet and outlet glaciers of the Vatnajökull ice cap in Iceland.

Three video stations marked three levels of the journey: glacial heights, descent, and ocean. One highlighted feature was sediment. From humble beginnings at the still bottom of cryoconite holes and drifting slowly down tiny rivulets flowing over ice, the journey of sediment and water powerfully culminates in explosive brown rivers pouring into the ocean—tinged by the collective sediment of these multitudinous tiny features now combined. I happened to capture a spike of particularly rapid melting, evident in surging rapids. Glaciers and ice sheets lose mass not only by melting but also by calving, when chunks of ice drop from sheer ice faces into seas below. Icebergs and ice chunks contrast in their purity of white and blue from these meltwater rivers.

Finally, I arranged an interdisciplinary panel discussion at the exhibit, titled *Direct Encounters with Arctic Landscapes*. The discussion featured scientists, humanists, and artists (including my hybrid self) who have spent a significant amount of time in the Arctic reflecting on its rapidly changing landscape and how that has impacted their professional and personal lives. Participants told stories of adventure, forging personal connections with Arctic landscapes, and the challenges of working in a rugged environment.

3.3.2 Raven's Light

The origins of this piece began over years of different journeys to Arctic Alaska in response to a call for student contributions to a collective exhibit, *Mapping Home/Collecting Truths: Works by Indigenous and International Artists* [19].

This was part of a larger initiative, Documenting Change: Our Climate (Past, Present, Future), a program that "presents how strategies of visualization and observation are shared by artistic and scientific disciplines to record, study, and communicate climate change." The works featured in the *Mapping Home* exhibit showcased "the types of scientific research and exploration taking place on the university campus regarding climate change and its effects on Indigenous populations." Based on this call, I sought to tell a story where I could bring together my research, issues of climate change, and Indigenous peoples.

My art would weave together photography and writing to convey these ideas. The collage is organized in a series of photo pairs [20]. The vertical picture at the center is the only exception, as it includes its own internal pair. Most sets are clear to the viewer based on their vicinity to one another within the collage, similar size, or similar visual content. Each represents a tension—between tradition and change, local and foreign, organic and industrial. My focus is on material culture, exploring the sacred embedded in everyday objects. People appear indirectly.

Every image is from Arctic Alaska. Most images are from the village of Anaktuvuk Pass, with a few from Nuiqsut, one from Allakaket, one from Fairbanks, and several from surrounding lands (including the Dalton Highway and Gates of the Arctic National Park). The only image I did not take is a map produced by BP (British Petroleum). The intended organization is roughly diagonal from upper left to lower right, with the images at these corners forming the first pair.

I was honored to have *Raven's Light* featured in an exhibit primarily featuring art by Indigenous peoples. As it was a creative interpretation by an outsider, it does raise questions of authenticity and representation. One way I sought to navigate this was to ensure I had proper permission and buy-in from the community in which I worked. This included an ethical review by the University of Colorado-Boulder's institutional review board, as well as direct engagement with the community, including tribal representatives. I oriented the piece around the people and places with whom, and where, I spent time, responding directly to these, as well as relevant academic and creative literature. Despite this, there is undoubtedly room for improvement in my process and products in the future.

3.4 Conclusion

Science, art, and Indigenous knowledge are three primary and powerful ways of witnessing Arctic environmental change. Each incorporates varying motivations, epistemologies, and histories. Each offers indispensable and irreducible perspectives. Yet each can also work in tandem with the others. There are, however, clear disparities between them, sometimes justified, but often not. Science is the most powerful, well-funded, and politically connected of the three, even as politicians far too often ignore or downplay scientific evidence. The value of Indigenous knowledge has gained increased recognition, particularly within the context of Arctic science. By contrast, art continues to occupy an ambiguous position within political, environmental, and academic discourse, often perceived as peripheral to "serious" inquiry.

Through my art, I have sought to not only convey environmental change but also deliberately interrogate and represent the intersections and departures between these approaches. Both *Sensing Ice* and *Raven's Light* explore this space of tension between these various forms of inquiry, but with different foci. The primary concern of *Sensing Ice* is the space between science and phenomenology (a mode of deliberate perception that seeks to focus on direct sensation free of or prior to scientific or metaphysical explanation; this intersects or shares elements with both art and Indigenous knowledge). The primary concern of *Raven's Light* is the tension between Indigenous knowledge and resource-oriented conceptualizations of place and landscape, the latter often closely interlinked with scientific presentation.

An underlying motivation for both projects was to convey an Arctic that transcends the objectified and scientifically transmuted. Climate change, for instance, is far more than a graph of ice loss, temperature measurements, or altered species regimes. The Arctic is a land of relations forged in everyday immersion in a sensual and living world.

These projects were in one way exploring universal concerns of knowledge and relationship with place and the nonhuman world. Simultaneously, they directly engaged with a rapidly changing world, documenting how traditional lifeways and places that have existed and been intact for millennia are being drastically altered or, in some cases, entirely erased.

Each project had a different focus: *Sensing Ice* on Arctic (and briefly Himalayan) ice in the context of rapid climatic change; *Raven's Light* on Arctic Indigenous people and the various changes shaping their lives and landscapes, though with intended resonance far beyond this limited context, engaging directly with viewers' lives and moral sense. The former highlighted the sensual and phenomenological, including my embodied experiences of Arctic places with an elemental focus; the latter centered on Arctic peoples, though somewhat indirectly through a focus on material culture. Each, however, told stories of Arctic change through art. With their disparate approaches and subject matters, they complement each other, together offering a richer and fuller portrait of the Arctic.

References

1. Hiraishi, K. *Exploring the Native Hawaiian Belief in the Sacredness of Mauna Kea (Internet)*. Honolulu, HI: Hawai'i Public Radio. August 5, 2019 (2025 Apr 3). Available From: https://www.hawaiipublicradio.org/local-news/2019-08-05/exploring-the-native-hawaiian-belief-in-the-sacredness-of-mauna-kea
2. Penn, I. and Elliot, R. *In U.S. Ramps Up Hunt for Uranium to End Reliance on Russia (Internet)*. New York, NY: The New York Times. Sep 30, 2024 (2025 Apr 3). Available From: https://www.nytimes.com/2024/09/30/business/energy-environment/uranium-russia-united-states.html

3. Bharath, D. and Molina, A. *Oak Flat is Sacred Land to Some Native Americans, But It's Endangered by a Plan for a Mine (Internet)*. New York, NY: The New York Times. June 28, 2023 (2025 Apr 3). Available From: https://apnews.com/article/oak-flat-sacred-site-arizona-copper-mine-230f572518858733462ac3fd6ca3be14)
4. Eaton, D. *Fighting for What's Sacred in the Arctic National Wildlife Refuge (Internet)*. San Francisco, CA: The Spiritual Edge (KALW Public Media). Sep 30, 2020 (2025 Apr 3). Available From: https://alaskawild.org/blog/fighting-for-whats-sacred-in-the-arctic-refuge/
5. Äikäs T. From boulders to fells: Sacred places in the Sámi ritual landscape. *Monographs of the Archaeological Society of Finland*. 2015; 5: 1–294.
6. Berkes F. *Sacred Ecology*. Routledge; 2017 Sep 1.
7. Borgmann A. *Real American Ethics: Taking Responsibility for Our Country*. University of Chicago Press; 2010 Jun 15.
8. Ballard C. The sun by night: Huli moral topography and myths of a time of darkness. Fluid Ontologies: myth, ritual and philosophy in the Highlands of Papua New Guinea. London, UK; 1998: 67–85.
9. Borgmann A. Broken symmetries: The romantic search for a moral cosmology. In *Philosophical Romanticism*. Routledge; 2006 Sep 27 (pp. 238–262).
10. Borgmann A. The force of wilderness within the ubiquity of cyberspace. *AI & Society*. 2017 May; 32: 261–265.
11. Lopez B. *Arctic Dreams: Imagination and Desire in a Northern Landscape*. Random House, New York, NY; 1999.
12. Abram D. *The Spell of the Sensuous: Perception and Language in a More-than-human World*. Vintage, New York, NY; 1997 Feb 25.
13. Gearheard S. *The Meaning of Ice: People and Sea Ice in Three Arctic Communities*. International Polar Institute Press, University Press of New England, Lebanon, New Hampshire; 2013.
14. Lopez B. *About this Life: Journeys on the Threshold of Memory*. Vintage; 2011 Sep 14.
15. Monroe, M. Chasing Ice (Film). Orlowski, J., producer. New York, NY: Submarine Deluxe. January 23, 2012. 75 Minutes.
16. Extreme Ice Survey (Internet). Available from: https://earthvisioninstitute.org/extreme-ice-survey/
17. Burtner M., biography (Internet). Anchorage, AK. Available from: https://www.matthewburtner.com/about
18. Sensing Ice. *A Philosophical-Aesthetic Investigation of Immediacy and Mediation (Internet)*. Publisher: Chris Dunn, Academia.edu. Published: 2024. Available From: https://www.academia.edu/119065931/Sensing_Ice_A_Philosophical_Aesthetic_Investigation_of_Immediacy_and_Mediation
19. *Mapping Home/Collecting Truths: Works by Indigenous and International Artists (Internet)*. Publisher: University of Colorado Libraries. October 28, 2018. Available From: https://libraries.colorado.edu/2018/10/08/mapping-home-collecting-truths-works-indigenous-and-international-artists
20. *Raven's Light* (Internet). Publisher: Chris Dunn, chrisdunnonplanetearth.com. Available From: https://www.chrisdunnonplanetearth.com/ravens-light.html

4 Sanaaq Siḷamiñ (Something Whittled from Weather)

Erin Ggaadimits Ivalu Gingrich

4.1 Introduction

To witness and observe is critical to knowing a place and it demands time, attention, and presence. The Indigenous peoples to which I belong, the Koyukon Dené and Iñupiaq, practice this in perpetuity in relation to the nuna and siḷa to which they belong.[1] Throughout Inuit Nunaat the practice of observation is central to Inuit ways of being.[2] In the essay "Curating Inuit Qaujimajatuqangit: Inuit Knowledge in the Qallunaat Art Museum" (2017), Inuk scholar Dr. Heather Igloliorte asserts observation as one of six tenets that guide Inuit ontologies in relating to other Arctic beings both human and non-human. These observational practices are ancestral, and it is the continued surviving and thriving of Inuit within their homelands that serves as the proof of these deep ancestral wells of witnessing.

This ancestral Iñupiaq practice of observation has not been without change. Adaptation is sacred and has always been key to our survival. These observational practices make knowledge that is then documented, recorded, and marked in many ways. The pictographic markings carved and etched into ivory drill bows depictively tell stories, histories, and hold hunting tallies as is addressed in "Our Stories Etched in Ivory: Qulip'yugut Iksiaqtuumaruat Tuugaami." The adaptation to forms of recording can also be seen within the work of Iñupiaq artist and illustrator James Kivetoruk Moses whose works in paper and pencil represent Iñupiaq life ways in color and details that require an intimate knowing. The adoption of introduced technologies also has changed how we record observations, like the use of the qiñiġaun, which was done by Iñupiaq teacher and photographer Charles Menadelook, whose distinct black and white photographs were made in the Northwest of Alaska from 1907 to 1932.[3] These practices all accumulate and adapt to make Iñupiaq observations tangible in new ways. During the 1996 Imaging the Arctic conference, keynote speaker Inuk Photographer George Quviq Qulaut stated, "Our ancestors and ourselves have been able to adapt to new surroundings and changes. Photography is no exception."

DOI: 10.4324/9781003545880-6

Right now, as changes are escalating within nuna and siḷa, these Indigenous practices of witnessing place and weather become increasingly critical, and it is through these processes of observation, documentation, and recording that wayfinding becomes possible. This writing is led by Iñupiaq Language as an anchoring framework that proposes an Indigenous wayfinding process utilizing the camera called: Qiñiġaaq iglauyaaq.[4] The Qiñiġaaq iglauyaaq methodology assists in tangible research processes to support the recognition of carving as a tangible discourse for addressing the climate in nunaat. Understanding Northern Indigenous relations to place and climate is supported by the reassertion of the Iñupiaq Language and cosmologies by representing our weather as breath with which to adorn carvings of wild relatives with images of weather and place.

Climate change is a felt experience that does not exclude representing it as held by non-human beings and land lay space for the plurality of experiencing the dysregulation of weather both externally and internally. The location of our siḷa within our body and lungs situates climate change within the animate being as an active phenomenon that we all carry. These are the places and land that make our bodies and the weather that we carry within our lungs. As innate beings of Inuit Nunaat, we share the burden of carrying an unstable breath and still we move forward in every direction (Figure 4.1).

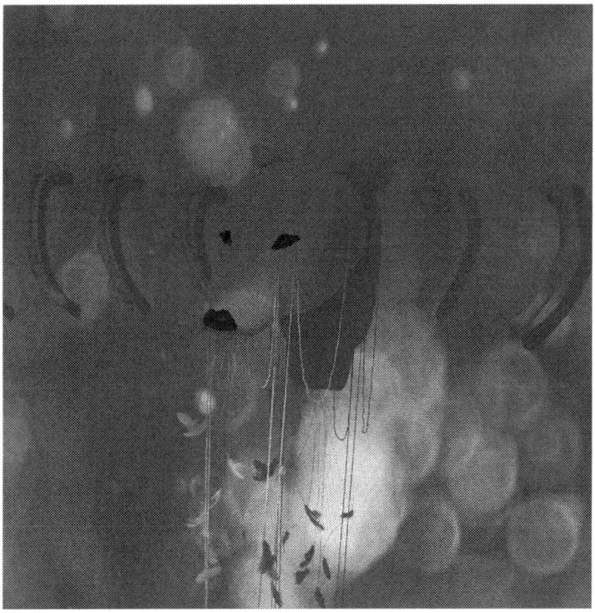

Figure 4.1 Nanuq aniġniq with nuvuliġaq from Utqiagvik, Carving installation 2024.

4.2 Siḷamiñ

Siḷamiñ stands as a representation and record of weather with new forms of adornments that have been created using the camera while still being centered on the first Iñupiaq ancestral forms of discourse with weather, which are found within the practice and histories of carving. This work prompts questions of proposing the possibility of a strategy for persistent covert resistance and defiance of colonial systems by utilizing the camera, or carvings, or language to navigate away from the dispossession of the homelands we belong to. Or maybe it is a call to answer the climate crisis we are experiencing that is removing the breath that is our weather from our homelands.

Some of the oldest and most prolific pieces of Iñupiaq material culture held in nuna are carved materials. Whether it is wood, ivory, antler, horn, stone, fossil, or bone, carved materials have survived and allowed us to survive in a nunaat in the Arctic that holds one of the coldest climates. These objects, which have their own being and autonomy, herein called object beings, are ancestors, Elders, teachers, and transmitters for our Iñupiaq cultural iconography, cosmology, visual language, and their associated frameworks, both awake and dormant. Iñupiaq carvings and masks have an ancestral purpose of being functional and a history of interacting with the intangible and non-human beings.

Iñupiaq carving serves as a discourse and method to create functional object beings in perpetuity. Holding an ability to address the climate disruptions we are experiencing here in the North when applied, Iñupiaq carvings can function to continue our ancestral dialogue with our homelands and weather. It is not enough to be using colonial and singular forms of knowing to understand and communicate the implications of our changing climate. The colonial structure invasively creates separation between Northern Indigenous peoples and their nunaat and siḷa, seeking to dispossess them of their nunaat. The renegotiation of language and translation can reaffirm frameworks of care and reassert Northern Indigenous epistemologies that facilitate the reciprocity that surrounds and founded the ancient relationships that are held with nuna and siḷa.

During this time of climate unbalance and trauma, arctic and subarctic bodies of the circumpolar North, whether land, human, or non-human, hold this unbalance in our being. The ecological dysfunction of nuna and siḷa is felt in our bodies and breath as both are given from the land and weather, respectively. This connection works bilaterally, and as our lands and weather change, so do we (Figure 4.2).

Weather is an active physiology of nuna. Climate dysfunction/dysregulation is causing literal and cultural displacement and straining the ancestral bonds to nunaat. The return of ancestral discourse with siḷa through carving practices, mask making, Iñupiaq Language, and their frameworks with contemporary adornments is an act of maintenance for the Indigenous inheritance of belonging to a homeland and belonging to a weather.

Nunaat's siḷa was active in protecting my ancestors from elements of colonial invasion. Arctic sea ice and weather played a critical role in limiting colonial agents from access to my ancestors and the homelands to which they belonged. This protection from siḷa gifted us with inherent environmental boundaries

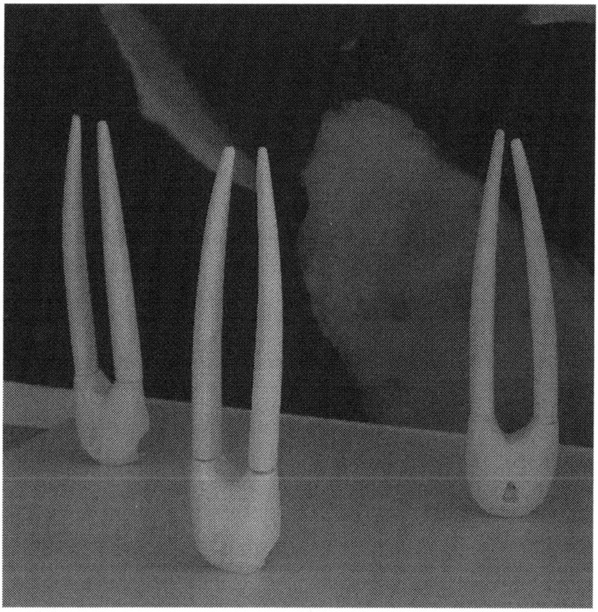

Figure 4.2 Aivġit auk from Sitŋasuaq, Carving installation 2024.

that prevented an easy arrival of most outsiders and have been an indispensable factor in my ancestors' and Elders' survival and resistance to the arrival of colonial oppression to nunaat. These bonds with weather and our homelands have remained majorly uninterrupted since memory, and it is in times such as now that they must be cared for with the frameworks and practices made to engage with them activated and threaded back.

The Iñupiaq have been carving perpetually, and it has always served as a functional way to make the things we need to survive. These object beings encode histories and inheritance as kiŋuvaat, both recognized and unrecognized.[5] Thus, as mentioned, Iñupiaq carving is a tangible ancestral discourse that exists as a site for function and active communication with the world around us.

Right now, siḷa is being witnessed and engaged with in many ways. Due to the destabilization of our relationship with siḷa, we hold so much to disclose, express, and communicate. Indigenous Peoples of the North have been using art as a form of wayfinding, discourse, and study of their nunaat and siḷa since the beginning of our orally recorded histories. Representing and mirroring what is seen, felt, and gathered echoes back to when our lands became homelands. Some of this work is within contemporary mediums, and others follow more ancestral discourses. Before continuing to address the weather, it is important to understand the Iñupiaq cosmology that it is directly tied to. The primary concept of this comes from my Iñupiaq Elder and Language professor.

Within my personal archive, notes from Iñupiaq classes taken in the early 2010s still serve in guiding me to how to view things in the Iñupiaq way. In 2011, at Iñupiaq class, my teacher described what happens when Iñupiaq die, the kinds of human spirits that there are, and the cycles of life, death, and being that are known by Iñupiaq ancestors and other relatives. My Elder described the three components of being and how they cycle in the world. The body returns to the nuna from which it came; this happens because the bodies are from the land, and they must return for the balance of the cycle to continue. The breath returns to siḷa; the origin of its balance is found there in the air, weather, and atmosphere from which it came. And the spirit returns to its name, which is given to the new life that enters during the same time as the old has departed. This is in relation to the atiq, which creates the bonds of kinship over entire communities, making us relatives to families of those whose name was gifted from and to. These understandings are known and unknown in many ways, but their presence provides a foundation from which to understand representations of weather and how it is tied to breath. This linear connection prompts representations of weather to be present within depictions of breath as they exist in balance, reciprocity, interdependence, and simultaneity.

Within Iñupiaq carving, representations of weather and breath already exist in forms and were designed to convey these elements and beings with presence. There are two primary examples of these elements represented in carving. These are visually referenced in the carved work and communicate the breath interacting with the atmosphere and environment as it does so heavily here (Figure 4.3).

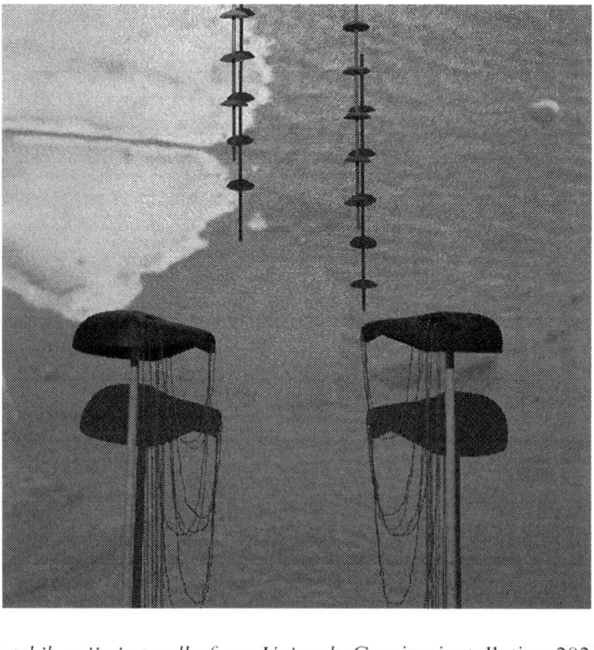

Figure 4.3 Natchik aniġniq to allu from Ugiuvak, Carving installation 2024.

The addition of projected and installation elements to envelop and adorn the carved work gives them a breath of weather. Photographs of homelands they are in relation to give them connection to the places where their breath is held. These new elements create adornment and activation that serve to echo the activation of masks through dance or performance as is the historical methodology to activate these carved forms. Adapting from the practice of dancing masks and carved works to the playing images of weather and place that dance on and around the masks. Siḷamiñ does not reject the act of dance and performance that was and is still critical to activating masks but instead moves laterally to create an adornment activation that performs the breath that is outside of the body as we know it in the weather.

These projected films and printed photographic adornments evoke the weather as it exists externally, while the carvings hold representations of the weather that is held internally in breath form. Siḷamiñ creates a weather that is not just to hold but that is also to be held. These marked times of change are adorned with tattoo forms that have marked me as I have leaned into my traditional tattoo journey, echoing forms adorned on body and land.

4.3 Conclusion

As an Indigenous person from the North, our histories tell me that we have always been witnessing the world, weather, and land changing around us, and we have decisively always been finding our way through it. Adaptation is one of our most sacred traditions and that has not changed but what we have adapted to is existing under a colonial structure that restricts the ancient relationship that we have with our homelands. We have ways of communicating, documenting, and witnessing the land and weather we belong to; it is time for them to be relearned and awakened to help us as we continue to find our way in our ever-changing nunaat.

Acknowledgment

This writing has been adapted from the author's 2024 MFA thesis paper titled: *Siḷamiñ: Timikput nunmiñ suli aniġnikput siḷamiñ.*

Notes

1 Nuna – land; siḷa – air, weather, and atmosphere.
2 Inuit Nunaat – Inuit homelands.
3 Qiñiġaun – camera.
4 Qiñiġaaq iglauyaaq – photograph path.
5 Kiŋuvaat – inheritance, knowledge from the ancestors.

Bibliography

Alaska Native Cultural Heritage and Artistic Sovereignty in Museums – the CIRI Foundation. (n.d.). https://thecirifoundation.org/project-grants/alaska-native-cultural-heritage-and-artistic-sovereignty-in-museums/

Hallendy, Norman. 2009. *Tukiliit: The Stone People who Live in the Wind: an Introduction to Inuksuit and Other Stone Figures of the North*. Douglas & McIntyre.

Igloliorte, Heather. 2017. *Curating Inuit Qaujimajatuqangit: Inuit Knowledge in the Qallunaat Art Museum* (2017).

Inuit Circumpolar Council. 2022. "Categorization is Political: Using Grounded Theory in the Creation of Controlled Vocabularies." https://iccalaska.org/. Accessed April 4, 2024. https://iccalaska.org/wp-icc/wp-content/uploads/2022/04/AOS-2022-Categorization-is-Political.pdf

King, Jonathan C. H., and Henrietta Lidchi. 1998. *Imaging the Arctic*. University of Washington Press.

Metchewais, Kimowan, Christopher Green, Emily Moazami, and Jef Whetstone. 2022. *Kimowan Metchewais: A Kind of Prayer*. Aperture Foundation.

Norbert, Eileen, ed. 2017. *Menadelook: An Inupiat Teacher's Photographs of Alaska Village Life, 1907–1932*. University of Washington Press.

Phillips-Chan, Amy, ed. 2021. *Our Stories Etched in Ivory*. Arctic Studies Center, Smithsonian Institution.

Shaginoff, Melissa, and Murphy Bess. 2022. "Conversations Of Ourselves – Conversations Of Ourselves." Coe Center. Accessed April 4, 2024. https://www.coeartscenter.org/conversationsofourselves/

University of Alaska Museum of the North. n.d. "Ethnology and History Collections, Mammals Collections, Ornithology Collections, Archeology Collections." Fairbanks, AK.

Unknown. 2020. "King Island Wolf Dance Documentary." YouTube.com. July 17. Accessed April 4, 2024. https://www.youtube.com/watch?v=DVfKpWaz2_Y

Part II

Art as Response

5 Exploring Perceptions of Northern Landscapes through Nature Photography

Esa Pekka Isomursu and Petri Hoppu

5.1 Introduction

Despite the increasing susceptibility of photographs to technical, artistic, social, political, and commercial manipulation, faith in the realism of the photographic image persists and is still very much grounded in the belief that a photograph is a mechanical reproduction of reality [1]. However, this perspective overlooks the complexities involved in photographic interpretation. While documentary-style photographs may seem objective due to their technical precision, they are shaped by various subjective factors such as subject selection, framing, and viewer biases. Moreover, perceiving a photograph does not correspond to the original sensory experience in physical reality, although they are related. Therefore, it is crucial to critically engage with the complexities of photographic representation, recognizing the active role of both producers and viewers in shaping meaning and interpretation, and appreciating photographs as multifaceted artifacts reflecting diverse influences.

Murray [2, p. 99] describes immersion as a metaphorical term derived from the physical experience of being submerged in water. We seek a similar sensation from a psychologically immersive experience: the sensation of being surrounded by an entirely different reality, as different as water is from air, that takes over all our attention. When different senses are engaged simultaneously, the likelihood of immersion may increase because it reinforces the feeling of presence or involvement. However, a multisensory experience does not always result in immersion: one can be exposed to a variety of sensory stimuli without feeling absorbed or engaged. Conversely, it is possible to feel deeply immersed through just one or two senses, such as in the case of a film (sight and sound) or a gripping audiobook (sound alone).

While a nature landscape photograph has the power to stir emotions and conjure feelings associated with being in nature, it cannot replicate the complete sensory immersion of being present in person. But to what extent can a photograph convey the photographer's experience? In this chapter, we explore the extent to which photographs of Northern landscapes can challenge ocularcentrism, evoke a holistic, multisensory experience of being in nature, and identify triggers for these responses. We also ponder the role and meaning

DOI: 10.4324/9781003545880-8

of authenticity in this context. Our investigation draws upon the results of a survey we conducted, interpreted in conjunction with insights gleaned from relevant literature.

The concept of landscape in Western culture carries strong connotations of an outsider's view, framed and seen from a distance [3, pp. 20–22]. For centuries, the North has been imagined and represented by Western artists and writers. Over time, the accumulation of successive layers of discourse has led to the creation of an 'Imagined North', as described by Chartier [4]. Although this Imagined North continues to thrive, particularly in tourism and on social media, local artists are increasingly seeking to deconstruct it. Landscape is no longer a genre solely tied to the depiction of nature; it has become a fluid concept, encompassing social, political, ecological, psychological, and technological dimensions, as exemplified by the works of Northern artists Marja Helander and Ella Morton, and more globally [5]. In this context, the concept of landscape closely aligns with that of the environment in environmental aesthetics. Berleant [6] emphasizes that the environment not only surrounds us, but that we, as experiencers, are sensorially immersed within it, forming a continuum with it. In posthumanism, the relationship between humans and nature is viewed as even more intertwined—and equal. We suggest that by gaining a better understanding of how photographs can convey this sense of immersion, our chapter provides new tools for artists to expand the expressive means of landscape photography.

5.2 Theoretical Foundations

Long before the permanent photograph was invented in the early 1800s, there was the camera obscura—a darkened room, or a box into which an exterior image was projected in real time via a small hole or a lens. Typically, the viewer would be located inside the camera obscura, making the viewing experience immersive and corporeal. The verisimilitude of the image was enhanced by the fact that the scene projected by the camera obscura was in its immediate vicinity. The viewer would most likely have observed this scene at some point or at least been aware of its nearby presence. However, one aspect that restricted immersion was that the projected image was upside down.

During the 17th and 18th centuries, the camera obscura was the most widely used model for explaining human vision, and for representing the relation of a perceiver and the position of a knowing subject to the external world [7, p. 27]. It stood as a model of how direct observation leads to truthful inferences about reality. Crary [7, pp. 39–41] argues that by introducing an exterior apparatus to create the observed image, the camera obscura separates the act of seeing from the physical body of the observer, thereby decorporealizing vision: 'The camera obscura a priori prevents the observer from seeing his or her position as part of the representation. The body then is a problem the camera could never solve except by marginalizing it into a phantom in order to establish a space of reason'. This concept aligns with the ideas of the homunculus fallacy and the

Cartesian theater, where an internal observer (a homunculus) views images formed on the retinas as if they were projected onto a movie screen [8].

Until the early 19th century, it was commonly believed that sensory perceptions occurred outside the conscious body, with the senses transmitting external reality to the perceiving homunculus. Crary [7, pp. 13–14] argues that, around 1810 to 1840, before photography, a variety of discourses and practices, e.g., in philosophy and science, made the camera obscura as an explanation of our observation obsolete, and the human body became an active producer of the optical experience. He gives the stereoscope and afterimages as examples of what led to this change. The early Wheatstone mirror stereoscope had a pair of mirrors at 45-degree angles to the user's eyes, each reflecting a picture located off to the side. As each eye could see only the image designed for it, it became obvious that the fusion of the sensory data from the images into a three-dimensional view had to happen within the observer. Also, Goethe's [9] research on afterimages showed that human perception can change without an external object of perception: an afterimage changes in time after the observer closes eyes, without any external stimulus being present [7, pp. 67–69].

Ontologically, the emergence of the permanent photograph has profoundly transformed image perception. Unlike the real-time image in a camera obscura, which requires the presence of the device, a permanent photograph freezes a moment in time which immediately becomes the past, rendering the image independent of the photographic device and its surroundings. However, an inherent connection to reality always remains in a photograph as well:

> I call "photographic referent" not the optionally real thing to which an image or a sign refers but the necessarily real thing which has been placed before the lens, without which there would be no photograph. Painting can feign reality without having seen it. (...) in Photography I can never deny that the thing has been there. There is a superimposition here: of reality and of the past. And since this constraint exists only for Photography, we must consider it, by reduction, as the very essence, the noeme of Photography.
>
> —Roland Barthes [10, pp. 76–77]

Seppänen [11, p. 72] refers to this connection between photographic representation and its objects, caused by photons, as a material trace. He suggests that this trace perpetually fuels epistemological controversy: when a photograph is perceived as a representation of an absent object, the object is still present as a material trace. Conversely, when the photograph is seen as a present material trace of the object, the photograph as re-presentation renders the object absent. Therefore, a photograph is always an unstable, unruly representation where presence and absence can alternate, much like the perceived image in a Rubin's vase [12]. Consequently, a photograph not only conventionally refers to something external to itself as a pictorial representation, but also the external reality can be considered present within the photograph itself [13, pp. 9–12].

In contemporary culture, owing in part to the ubiquity of digital media, visuality has become accentuated at the expense of other sensory modalities, particularly sound and touch. Visual perception has emerged as the predominant means through which individuals comprehend their surroundings and process information, thereby dominating the discourse. The diversity of visual communication, encompassing images, symbols, and graphics, provides a broad spectrum of expressive possibilities. Notably, the sense of touch had been an integral part of classical theories of vision in the 17th and 18th centuries, but it was dissociated from sight in the 19th century. The loss of touch as a conceptual component of vision meant the isolation of vision from tactility and its subjective relation to perceived space [7, p. 19]. Furthermore, the emphasis on minimizing physical contact during the COVID-19 era led to the sense of touch becoming pathologized [14]. Nevertheless, this does not imply an inherent inferiority of other senses in the holistic human perception of the world.

Bowring [15] states that ocularcentrism, or the hegemony of the eye, has dominated the Western paradigm of engagement with the world since the Renaissance. This is illustrated in the term 'visual culture' which encapsulates the breadth of practices associated with the eye, the gaze, and the operations of contemporary culture—everything from digital, to cinema, to photography, and the consumption of all things visual. Bowring sees the dominance of the visual as a double-edged sword. On the one hand, visuality is aligned with knowledge, and there is a plethora of figures of speech in the English language which connect understanding with sight, 'I see' being one of the most obvious. On the other hand, however, pure opticality can also be interpreted as a failure to fully engage, an 'overlooking' of the richness of ideas that might be within what one beholds. The alignment of the 'eye' and the 'I' emphasizes the distancing of the subject from the object, the 'disembodied eye' [16, p. 81], or in terms of landscape, a detachment of the self from place.

Parting from the Cartesian dualism of mind and body, Merleau-Ponty [17] emphasizes the body as the primary site of knowing the world, instead of the consciousness or the mind: a painter paints with his body, and a viewer views with his body. Merleau-Ponty [17] compares the body to the heart in an organism: it gives life to the inner spectacle that we experience as a world. Following Merleau-Ponty, Hoppu [18, p. 43] states that '(o)ne's relationship to the world (as a body) is to a great extent non-conceptual, something that cannot be articulated with propositions'. What we regard as knowing is not merely related to our cognition or language, but knowledge consists of our embodied activities as well [19, pp. 151–154].

In his essay 'Cézanne's Doubt', Merleau-Ponty [20] finds Cézanne's theory of painting analogous to his own philosophical thought. In Cézanne's series of impressionist oil paintings of Mont Sainte-Victoire, he not only shows the landscape but also leaves visible the elements that make up the painting. Traces of brushstrokes are easily recognizable, and he has left patches of the canvas uncovered by paint. The impression of these paintings is twofold: while contemplating the painting, one can concentrate on the subject matter, i.e., the landscape, or one can concentrate on the surface structure, the paint, and the pattern of colors. The spatial depth of the painting does not correspond to that of the real world but instead follows 'the lived perspective', that which we actually perceive, not a geometric or photographic one. This lived perspective is rooted in an experience prior to the artificial isolation of the senses and the hegemony of sight. Thus, Cézanne tried to present objects that were present to all the senses at once, claiming that we could see the depth, the smoothness, the softness, the hardness, and even the odor of objects [20, p. 14]. Cézanne's comprehension that expression comes from the material, from the paint itself, defined Merleau-Ponty's idea of artistic expression as well. It is not only the subject matter of the painting, but also all these elements together, corporeally experienced by the viewer, that convey the artist's vision.

On a similar note, Shusterman [21] argues that we tend to identify photographic art so one-sidedly with the end products, i.e., the photographs themselves, that it occludes the aesthetics arising from the somatic, performative process of taking the photograph. Although the photograph is the standard end product of photography and conventionally recognized as the goal of photographic art, the process itself of making—not merely 'taking'—a photograph can constitute a real aesthetic experience and value that cannot be found by contemplating the mere end product. In our chapter, as explained later, we sought to determine whether making this process visible could also help the viewer more closely relate to the photographer, thereby enhancing immersion.

In addition to the viewers' preoccupation with the photograph as an object, Shusterman [21] gives other reasons for the collective negligence in the process of photography. The fact that pressing the release of the camera shutter requires no special skill or thought and that the camera mechanism automatically does all the rest to produce a realistic photograph diminishes the sense of photography as a performative achievement. The instantaneous act of shutter release likewise suggests that there is no sustained duration of effort involved as one would expect in a performative process. These reasons, however, neglect the complex performative process executed by the photographer that occurs before the shutter release.

The performative process of photography includes all the actions that are necessary for achieving the desired optical image that then forms the photographic image on film or a camera sensor. This includes, among other things, composing the shot in one's mind, selecting the right lens and exposure values, moving to the right position, and possibly staging the scene as well as guiding the model(s), or posing as a model. In a wider sense, this can be thought to include all the previous work that the photographer has done to gain the skills s/he has as a photographer. The fact that this whole process is ontologically complex and difficult to demarcate in its experiential dimensions provides further reasons for its neglect [21].

In the context of the perception of photographs, authenticity often relates to the ability of photographs to convey or capture reality in a certain way. By using photographs as a document or record of a specific moment or event, people may perceive them as offering genuine glimpses into the past or other places and events. In this sense, the authenticity of photographs relates to their ability to convey reality or at least a perspective of it.

Authenticity as an experiential value is a Modernist construction, first articulated in Rousseau's Romantic philosophy [22, p. 38]. The continued craving for experiences of unmediated genuineness seeks to cut through what Rousseau called 'the wound of reflection', a reaction to modernization's demythologization, detraditionalization, and disenchantment. The notion of authenticity implies the existence of its opposite, the fake, and this dichotomous construct is at the heart of what makes authenticity problematic. Identifying some cultural expressions or artifacts as authentic, genuine, trustworthy, or legitimate simultaneously implies that other manifestations are fake, spurious, and even illegitimate. Disciplinary practice has 'nostalgize(d) the homogeneous' and decried 'bastard traditions', thus continually upholding the fallacy that cultural purity rather than hybridity is the norm. Today, in practice, the genuine and the spurious are often converging, their identities separable only by their narratives [23, p. 9; 24, pp. 8–9].

Benjamin's [25] concept of aura can be seen as adding depth to the concept of authenticity in photographs. Benjamin famously argued that an original artwork possesses a certain quality he called aura, a sense of 'here and now', that is inherent to its singularity. This aura is associated with the time and place of the artwork's creation, as well as its physical presence. The aura is closely tied to the uniqueness and authenticity of an original work of art and is easily lost in the age of mechanical reproduction. This is especially strongly manifested in photography and film.

Groys [26], however, argues that, although the idea is often neglected in the interpretations of Benjamin's work, Benjamin also opened up the possibility of reproductions becoming auratic, original pieces. Groys interprets Benjamin's concept of aura as purely topological. The original has its proper place, an aura: the copy is out of place and without history and therefore without an aura. The difference is decided by the observer: the original is the one you go to, and the copy is the one that comes to you. Groys compares the Internet surfer to Benjamin's flâneur. The flâneur does not demand objects to come to him but goes to them and thus encounters them as 'original'. Similarly, in their mind, Internet surfers visit the 'original' web addresses rather than ordering the content to the terminal they are using, even though the latter is technically the case. A key feature of the Internet is that all signs, words, and images are assigned a unique address, i.e., they are localized, territorialized, and written into the topology. In this sense, the information network makes an original of each record that may have originally been created as a multiple copy [26–28].

Groys extends the concept of the flâneur to Internet users, following an academic trend emerging in the post-1990s with the World Wide Web. As flânerie acquired new meanings on the Internet, new terms were derived, such as cyberflâneur [29], virtual flâneur [30], and online flâneur [31]. We use the term cyberflâneur for all Internet-based flânerie.

As Benjamin [32, p. 62] points out, an active flâneur who enjoys the power of observations and is dominated by curiosity may easily turn into a passive badaud, a gaper dominated by sentiment. Echoing Benjamin, Badami [33, p. 93] states that there is a thin line between a cyberflâneur and cyberbadaud. The ideal has been a curious wanderer that seeks authentic knowledge, explorations, and connections in the limitless virtual world, but the reality is often 'mindless mob mentality of (cyber)badauderie', demonstrated by 'frictionless sharing of cat videos and retweeting meaningless memes on Facebook and Twitter, users thoughtlessly clicking likes or re-sharing others' recycled links, the hysterical screeches of outrage and maudlin sentimentality, the credulous posting of hoaxes and celebrity gossip, (and) the dependently narcissistic anxiety over likes and friends and followers(.)' [33]

Thus, being a challenged concept even before the Internet revolution, authenticity has increasingly become associated with its experiential value—a development that began already in the early modern era. Cyberflâneurs seek to establish their experiences as genuine even in virtual reality, distinguishing themselves from cyberbadauds that are perceived as lacking the discernment to identify authentic web content. As Woolfe [34] suggests, 'it is not mindless novelty that the flâneur wants but new directions that feel organic and authentic'.

5.3 Survey Settings and Selection of Images

Based on the theories presented above, we curated twenty-one photographs (Figure 5.1) taken by the first author in northern Finland and Norway. This selection was used for a nationwide survey in Finland. The aim of this survey was twofold: to investigate whether landscape photographs depicting Northern nature can transcend ocularcentrism to foster multisensory immersion in viewers, and to identify specific subjects within these images that facilitate such experiences. The exact location depicted in the photographs—beyond being from the North—was not of significance. What was essential was that each photograph illustrated one or more of the theoretical perspectives we wanted to explore.

We selected photographs that were not overly distinctive within their genre to ensure respondents' focus on the survey questions rather than the image quality, deliberately retaining certain imperfections, such as tilted horizons, as part of our methodology. The selection included documentary-style images as well as images employing creative techniques.

Based on the ideas of Merleau-Ponty, Cézanne, and Shusterman, we hypothesized that incorporating visible traces of the creation process into images could enhance viewer immersion. Therefore, we selected images featuring various creative techniques, including multi-exposure, long exposure, intentional camera movement (ICM), high-ISO sensor noise, and infrared imaging. We also thought that making visible the embodied, active engagement of the photographer with the landscape could reveal the somatic, performative process described by Shusterman. This approach aimed at conveying the photographer's corporeal presence in the landscape, thereby enhancing viewer immersion and appreciation of the image.

Figure 5.1 Photographs used in the survey. $N = 141$. Photos: Esa Pekka Isomursu.

Some images included people. We theorized that the presence of a human figure could prompt viewers to consider the photographic process and the photographer's involvement during the shoot. To maintain consistency in the reference images, certain images were edited to remove human subjects.

The survey, conducted online via Webropol, was distributed nationwide through both national and local photography organizations and social media. A total of 141 respondents participated, though not all participants answered every question. Participants were asked about their age, photography skill level, and interest in nature photography, as these factors might influence their opinions. The six-step rating scale required respondents to rate themselves as either below or above average. Notably, 50% of the respondents were aged over 55 years. Furthermore, 71% of participants expressed a high level of interest in nature photography (rating their interest as 4 or 5 on a scale from 0 to 5), and 77% considered their photography skills to be above average. Gender was not specified, as it was deemed irrelevant to the study's objectives.

Each image was assessed with four identical five-point Likert scale questions (from Strongly Disagree to Strongly Agree) focusing on viewer immersion: 'When I look at the image: I perceive myself within the world of the image; I feel I can touch the world of the image; I feel the presence of a photographer; the image conveys an authentic experience of nature'. Subsequently, participants were prompted to rate the activation of their senses (sight, hearing, smell, taste, touch, and movement) using the same five-point Likert scale. Thirdly, a seven-point scale was utilized for four semantic differentials (Unemotional—Emotional; Static—Dynamic; Cold—Warm; Uninteresting—Interesting). Finally, respondents could also provide written comments on each image.

After answering questions about individual images, participants were asked: 'Do you have any additional thoughts whatsoever related to the topic?' They were also given the option to provide contact information for a potential follow-up interview.

5.4 Survey Results and Reflections

The survey results indicate that the images were indeed perceived in a multisensory and immersive manner. However, our theoretical assumptions were not fully confirmed. Elements we thought would trigger specific responses did not always do so. Additionally, experienced photographers (those with self-assessed skills above average), though not asked, provided written judgments on the images, offering unexpected insights into landscape photograph perception and contemplation.

Approximately 39% of the written answers (155 out of 399) appraised the quality of the images, offering straightforward judgments on whether the images were good or bad, or providing normative criticism regarding the image's content, composition, or technical aspects. Notably, only 5% of these evaluations originated from individuals with below-average skill levels, despite them constituting nearly a quarter of the respondents. In many cases, preconceptions appear to strongly influence perception, which means that judgements about the quality of an image supersede the actual viewing experience. This phenomenon can be examined through Husserl's [35] concept of 'natural attitude' and Scholes's [36] 'protocols of reading'. Husserl [35, p. 61] called the prejudices and preconceptions that dominate the perception natural attitude (natürliche Einstellung), referring to unconscious assumptions in everyday life. These are acquired through, and influenced by, specific sociocultural practices [37]. In other words, we can state that depending on their prior epistemological stances, viewers employ various reading protocols to make sense of the photographs [1, p. 106]. The protocols of reading are a formulation by Scholes [36], and they point to understanding, interpretation, and criticism when reading books, bodies like in dance [38, p. 5], or, as in this case, photographs.

Hence, it appears that individuals deeply engaged in photography tend to align their responses with the established conventions of the field, rather than drawing solely from their immediate experiences. Strong preconceptions might solidify one's stance toward perceived objects, directing focus to details appreciated within the specific sociocultural context. In contrast, inexperienced photographers, lacking such context, approach survey questions with greater openness.

Our pre-assumption that leaving visible traces of the creation process might increase the viewer's immersion in the image was refuted. The use of creative techniques, adding visible noise, or a person in the image did not have such an effect in our survey. This finding contradicts the ideas of Merleau-Ponty and Cézanne [20] and Shusterman [21] about the importance of leaving marks of the artist's creation process visible. A possible explanation for this discrepancy is the deeply rooted idea of photographic authenticity. Unlike a painting, which is not expected to contain a material trace [11], a photograph is often subject to expectations of authenticity. This is especially true for nature photographs, which are typically expected to be documentary and authentic depictions of nature [39], as can be seen from these answers to the survey:

> Hares do not jump in the sky!
>
> Excessive image editing is disturbing, at least if you think you are looking at a photograph. Editing changes a photograph into something else.

When a person was featured in the photograph, the presence of the photographer was perceived as more prominent. However, contrary to our initial assumptions, this heightened presence did not contribute to an increased level of immersion in or appreciation of the image. Again, this might be due to the differences in how people interpret the authenticity of a painting and a photograph.

The results also show that creative techniques used in photography strongly divide opinions. There are many more respondents considering photographs with creative techniques to be strongly uninteresting or strongly interesting compared to their documentary-style counterparts (Figure 5.2). This observation

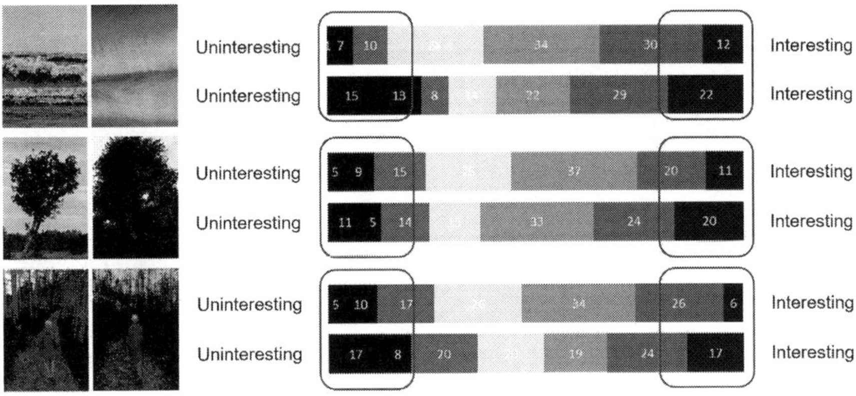

Figure 5.2 A comparison of images with documentary style and creative techniques.

aligns with our findings from an earlier survey [39], indicating a nearly equal division of people expressing support and opposition to creative techniques. This strongly polarized stance was particularly notable among respondents aged over 40.

The respondents often reported that, in addition to sight, some of their other senses were also activated when looking at the given photographs. The degree of activation varied among images and senses. The strongest multisensory response occurred with three documentary-style photographs of the sea (around the upper right corner in Figure 5.2). For these photographs, over half of the respondents reported experiencing activation in hearing and movement, and 43–57% in touch.

The sense of movement was also perceived by many images depicting a path, and images with flowing water (upper left and lower right, respectively, in Figure 5.2). Taste and smell were the least activated senses, with some activation mainly in documentary-style photographs of a stormy sea. Notably, photographs utilizing creative techniques cause significantly less activation in all senses. Thus, in the survey, a documentary style emerged as a prerequisite for a strong multisensory experience. An interesting anomaly was that nearly half of the respondents felt their sense of touch activated in response to the photograph with snow-capped trees and blue sky (middle left in Figure 5.2). This may be because the weather in the image appears particularly cold, evoking a sensation of 'biting cold', a condition very familiar to the Finnish respondents.

The photograph depicting a boy holding his head amid a storm generated a weaker sensory response compared to the version without the boy. This pattern was consistent across all images that had versions with or without a person present. As one respondent noted:

> (…) the person in the picture immediately distances me—I feel more like I am in the world of the picture if there are no others in it.

However, another respondent pointed out that image composition and the position of the person also have an effect. For example, the person walking in the middle of the image (see Figure 5.1) and the boy obstructing the view of the stormy sea block the viewer's ability to 'step inside'. In contrast, the photograph where a person is viewing a mountain lake from the lower left corner invites the viewer into the image.

The presence of a human figure in an image enhances its emotional impact. The image with a boy amid a storm, in particular, was perceived as much more emotional than the same storm image with the person edited out. This phenomenon, evident in today's selfie culture, can be traced far back in the history of the visual arts. In the first European drawing of Niagara Falls from 1698 [3, p. 136], the sublimity of the scene was emphasized by depicting individuals with hands in the air or covering their ears.

Approximately 9% of the written responses indicated that a person's previous experiences with nature strongly influence and extend their perception of the photographs:

> Makes you long to return to the autumn nature trail for a walk with the dog.
>
> I connect strongly with my past memories/experiences.
>
> I immediately remember the drops of the wave on the skin.

The concept of kinesthetic empathy [40] can explain this phenomenon. The more the viewer has experiences of movement in nature, such as hiking, the stronger their kinesthetic empathy and ability to understand the photographer's experience as a physical sensation. In another publication [41], we have further explored how familiarity with the depicted landscape—though not a prerequisite—can deepen the viewing experience.

In addition to activating more senses, documentary-style photographs were also perceived as more interesting. Their popularity compared to those employing creative techniques may, in part, be explained by the concept of aura [21, 25, 26, 28]. We posit that contemplating a documentary-style landscape photograph can be an auratic experience, much more so than with most other types of photographs. Aura, in this context, conveys feelings of uniqueness and authentic experience to the viewer.

Benjamin [25] identified two instances where photographs retained art's auratic cult value: pictures of loved ones, absent or deceased, and early portrait photographs where the lengthy process of taking a photograph created a ritualistic experience. The aura associated with the latter case diminished as the photographic process became faster, more commercial, and impersonal. However, Shusterman [21] argues that many ritual events, such as weddings, baptisms, and graduations, preserve some of the ritualistic elements of early photography, thereby retaining their aura and cult value.

In nature photography, there still remains a strong sense of ritual. The photographs serve as a testament to the photographer's journey into nature. Such trips always demand an—often substantial—effort and can be physically and temporally demanding. Thus, photographing nature involves a ritualistic transition to, and presence in, a meaningful location. There is an inherent presence of aura in landscape photographs compared to studio portraiture which often lacks a specific sense of place and fails to establish a topological connection. In contrast, Figure 5.3 evokes a strong sense of a unique location even in viewers unfamiliar with the exact setting, thereby creating an aura.

Groys's interpretation of Benjamin's aura as purely topological raises intriguing questions when applied to nature photography. As images generated by Artificial Intelligence, along with other manipulated images, become increasingly prevalent and difficult to distinguish from authentic photographs, ensuring authenticity becomes paramount for preserving the aura. We posit that the bond forged between the photographer and the viewer through their shared experiences and appreciation of nature imbues the photograph with its aura—but this aura will diminish if there is uncertainty about the authenticity of the photograph, thus compromising the authenticity of the original nature experience. We anticipate that, in the near future, the topological dimensions of aura will assume even greater significance in the perception of images as authentic. For instance, images showcased on esteemed museum websites or the social media accounts of recognized nature photographers possess a heightened aura

Figure 5.3 A nature landscape. Photo: Esa Pekka Isomursu (2015).

and sense of authenticity compared to those encountered randomly in social media feeds lacking a valued topological anchor point.

From the judgmental and analytical survey responses, we can clearly identify the cyberflâneur who contemplates the image, appreciating its aura and authenticity. From other responses, we can also identify the cyberbadaud, a passive gaper dominated by sentiment who is uncritical and interested only in the surface of the image: approximately 20% of written responses merely described the content of the image without further analysis.

> The beauty of Riisitunturi fell
>
> Lovely seascape by the coastline (...)

The cyberflâneur and the cyberbadaud can be seen as two ideal types [42] of the viewers. The cyberflâneur seeks realism—whether in documentary or creative forms—in the photographic portrayal of Northern nature. Conversely, the cyberbadaud is content with Baudrillard's simulacra [23], images that depict an imagined, superficially beautiful North but lack deeper meaning and authenticity—at least for the observer.

One might expect that environmental concerns would prominently feature in the open-ended responses to our survey, especially given the current global focus on such issues. However, out of the 399 written responses, not a single one expressed concern or took a stance on environmental matters, even though some of the images particularly encouraged it. Paradoxically, these images of nature may represent not only a return to the natural world but also an escape from reality. People seem to seek an idyllic connection with nature through these images, rather than confronting the harsh realities of environmental change. The survey responses reflect an idealized vision of nature, rather than acknowledging a changing and endangered environment. Perhaps this suggests that Chartier's concept of the Imagined North remains very much alive and well.

5.5 Conclusion and Future Work

Our study provides insights into how people perceive Northern landscapes through nature photography. It indicates that nature landscape images can indeed evoke a multisensory and somewhat immersive experience. The viewer's previous experiences with Northern nature, as well as the objects in the image and their composition, all influence immersion. However, for experienced photographers, previously acquired conventions and discourses of photography—i.e., protocols of reading photographs—dominate their viewing experience. This dominance may make it difficult for them to focus on aspects beyond what is formally expected from a 'classic', documentary-style, and traditionally composed nature photograph.

Drawing from Groys's topological interpretation of Benjamin's concept of aura, we argue that contemplating a documentary-style landscape photograph can be an auratic experience for the viewer, more so than with most other types of photographs. This aura brings feelings of uniqueness and authentic experience to the viewer. Protocols of reading, kinesthetic empathy, and the topological aura all explain why traditional, documentary-style photographs of nature are so popular. However, there is a notable minority, especially among more experienced photographers, who also perceive images utilizing creative techniques as authentic depictions of the experience of being in nature. Although we found a correlation between photography skills and an interest in creative techniques, our survey did not reveal the root cause behind this interest, if there is any. It would be intriguing to investigate further, for example, whether users of creative photography techniques possess a stronger-than-average knowledge of art history or have a background in other arts. We speculate that these characteristics might override the otherwise dominant 'classic' reading protocols of photographs. Furthermore, with the emergence of advanced image manipulation techniques, artificial intelligence, and the proliferation of simulacra, could the growing interest in and acceptance of creative techniques signal a paradigmatic shift? Such a shift might influence how Northern landscapes are perceived in the future.

The respondents' perception of the photographs can also be interpreted by the two ideal types that we identified, namely, the cyberflâneur and the cyberbadaud. The first enjoys the intelligent contemplation of the images, whereas the latter takes them at face value. Times have changed since Sontag [43, p. 55] described the photographer as a flâneur armed with a camera. Contemporary culture has created space for photographers—and viewers—akin to the badaud. There is now a significantly larger number of photography enthusiasts at all levels, as practically everyone carries at least a camera phone, and taking pictures incurs minimal cost. Had Sontag lived today, she would likely have acknowledged the role of the (cyber) badaud alongside the (cyber) flâneur among both photographers and consumers of photographs in today's visually saturated environment.

Many respondents have a strong personal connection with nature. Also, the increasing awareness of environmental issues and climate change, combined with post-humanist thinking, has fostered a sense of interconnectedness between humans and nature. This shift suggests that fewer people now view nature as something separate from themselves; there is a growing awareness of our dependency on nature—and nature's dependency on us.

The cyberflâneur's intellectual approach to observing photography and nature provides an opportunity to understand change and recognize that something can be done about it. In contrast, the cyberbadaud seeks momentary pleasure and remains indifferent to the state of things. Since these are ideal types rather than distinct individuals, elements of both perspectives can coexist in a single person's thinking. A potential area for future research is exploring how the mindset of the cyberbadaud can be shifted toward that of the cyberflâneur. Such a shift could enhance awareness and potentially inspire action to address environmental threats.

Our findings underscore the influence, and perhaps the responsibility, of both viewers and photographers in shaping how Northern nature is perceived and understood. By understanding the mechanisms behind the viewing experience, individuals can become more aware of their impact and better equipped to act responsibly toward the environment.

References

[1] Lévesque S, Ng-A-Fook N, Corrigan J. What does the eye see? Reading online primary source photographs in history. *CITE*. 2014;14(2):101–140.
[2] Murray JH. *Hamlet on the Holodeck: The Future of Narrative in Cyberspace*. Cambridge, MA: MIT Press; 1998.
[3] Andrews M. *Landscape and Western Art*. New York: Oxford University Press; 1999.
[4] Chartier D. *What is the Imagined North? Ethical Principles*. Montréal, Canada: l'Université du Québec; 2018.
[5] Davis H, Turpin E (eds). *Art in the Anthropocene: Encounters Among Aesthetics, Politics, Environments and Epistemologies*. London: Open Humanities Press; 2015.
[6] Berleant A. *The Aesthetics of Environment*. Philadelphia, PA: Temple University Press; 1992.
[7] Crary J. *Techniques of the Observer*. Cambridge: MIT Press; 1992.
[8] Dennett D. *Consciousness Explained*. Boston, MA: Back Bay Books; 1992.
[9] Goethe JW von. *Theory of colors*. Orig. 1810. London & New York: Routledge; 2019.
[10] Barthes R. *Camera Lucida–Reflections on photography*. Orig. 1980. New York: Hill & Wang; 1981.
[11] Seppänen J. *Levoton valokuva*. Tampere: Vastapaino; 2014.
[12] Rubin E. *Synsoplevede figurer: Studier i psykologisk analyze*. Copenhagen: Gyldendal; 1915.
[13] Seppänen J. Valokuva, materiaalisuus, representaatio. *Me&vi*. 2014;37(2):4–21.
[14] Burleson M, Roberts N, Munson A, Duncan C, Randall A, Ha T, et al. Feeling the absence of touch: Distancing, distress, regulation, and relationships in the context of COVID-19. *JSPR*. 2022;39(1):56–79.
[15] Bowring J. Sensory deprivation: Globalisation and the phenomenology of landscape architecture. In: Stewart G. editor. *Globalisation and Landscape Architecture*, 3–6 June 2007, St Petersburg: St. Petersburg State Polytechnic University Publishing House. 2007:81–84.
[16] Jay M. *Downcast Eyes: The Denigration of Vision in Twentieth-century French Thought*. Berkeley: University of California; 1994.
[17] Merleau-Ponty M. *Phenomenology of Perception*. Orig. 1945. New York: Routledge; 1962.
[18] Hoppu P. Other flesh. *NJD*. 2013;4(2):32–45.
[19] Hoppu P. Encounters in dance and music: Fieldwork and embodiment. In: Fiskvik A, Stranden M. editors. *(Re)Searching the Field –Festschrift in Honour of Egil Bakka*. Trondheim: Fakbolaget; 2014:151–159.
[20] Merleau-Ponty M. Cézanne's doubt. In: *Sense and non-sense* (H. L. Dreyfus & P. A. Dreyfus, trans.). Evanston, IL: Northwestern University; 1964. pp. 9–24.
[21] Shusterman R. Photography as performative process. *JAAR*. 2012;70(1):67–77.
[22] Richardson T. The peculiar deviance of authenticity. *Proteus*. 2014;30(1):31–36.
[23] Baudrillard J. *Simulacra & simulacrum*. Orig. 1981. Ann Arbor, MI: University of Michigan; 1994.

[24] Bendix R. *In search of authenticity: The formation of folklore studies*. Madison: University of Wisconsin; 1997.
[25] Benjamin W. The work of art in the age of mechanical reproduction. Orig. 1935. In: Arendt H. editor. *Illuminations*. New York: Schocken; 1969.
[26] Groys B. Auran synty: Muunnelmia eräästä Walter Benjaminin teemasta. Orig. 2000. *Nuori Voima*. 2002;4(5):58–61.
[27] Hanssen B. editor. *Walter Benjamin and the Arcades Project*. London & New York: Bloomsbury; 2006.
[28] Groys B. The aura of profane enlightenment. In: E. McDonald (Ed.). *The World (maybe) fantastic, exhibition catalogue*, Sydney: Biennale of Sydney; 2002.
[29] Goldate S. The cyberflaneur: Spaces and places on the internet. *Art Monthly*. 1997;208:15–18.
[30] Featherstone M. Virtual flâneur. *Urban Studies*. 1998;35:909–925.
[31] Hartmann M. Technologies and Utopias: The Cyberflâneur and the Experience of Being Online. PhD thesis, University of Westminster; 2004.
[32] Benjamin W. *Charles Baudelaire: A Lyric Poet in the Era of High Capitalism*. London: Verso; 1983.
[33] Badami S. The boulevardier of the information superhighway: Disconnection and discovery in the virtual global village. *Westerly Online*. 2016;1:89–99.
[34] Woolfe S. As in the streets, so in the mind: Flânerie as a way of thinking and living. 25 September 2022. (Accessed 26 June 2024). https://www.samwoolfe.com/2022/09/flanerie-a-way-of-thinking-and-living.html
[35] Husserl E. *Ideen zu einer reinen Phänomenologie und phänomenologischen Philosophie. Erstes Buch 1. & 2. Halbband. Husserliana. Edmund Husserl Gesammelte Werke*. Den Haag: Martinus Nijhoff; 1976.
[36] Scholes R. *Protocols of Reading*. New Haven, CT: Yale University Press; 1989.
[37] Weiss G. De-naturalizing the natural attitude: A Husserlian legacy to social phenomenology. *JPP*. 2016;47(1):1–16.
[38] Hamera J. *Dancing Communities: Performance, Difference and Connection in the Global City*. London: Palgrave Macmillan; 2007.
[39] Isomursu EP, Jokela T. Truthfulness in nature landscape photography. In: Coutts G, Jokela T. editors. *Relate North: Possible Futures*. Viseu: InSEA; 2023. pp. 74–97.
[40] Parviainen J. Kinaesthetic empathy. *Dialogue and Universalism*. 2003;11–12: 151–162.
[41] Isomursu EP. Visual Haiku: Haiku in the age of global visual media. In: G. Sade, A. R. Brown, L. Barclay, J. Seevinck, A. Tyurina, & R. Wright (Eds.), *Proceedings of the 29th International Symposium on Electronic Arts: ISEA2024 Everywhen. Volume 1: Academic Papers* (pp. 364–371). Kelvin Grove, QLD: QUT ePrints / Queensland University of Technology; 21–29 June 2024.
[42] Weber M. Die 'Objektivität' sozialwissenschaftlicher und sozialpolitischer Erkenntnis. *ASS*. 1904;19(1):22–87.
[43] Sontag S. *On Photography*. New York: FSG; 1977.

6 Nomadic Antlers

New Genre Arctic Art Education and Activism

Mirja Hiltunen and Korinna Korsström-Magga

6.1 Introduction

A most urgent and disastrous change in the Arctic is climate change caused by human exploitation of nature. This chapter follows the nomadic journey of the antler workshop in northern Scandinavia, the northernmost part of Europe. The same workshop was conducted in several places, inviting and encouraging youth in the Arctic to share their desires and thoughts for a sustainable future and express their concerns regarding changes in their everyday environment. This chapter emphasizes the importance of youth participation in climate action and highlights the need for context-sensitive approaches to climate activism and engagement among Arctic youth.

Young generations living in the Northern and Arctic regions encounter specific challenges related to climate change, which are heightened by their geographical isolation and socioeconomic conditions. They experience the impacts of climate change directly because the environment is changing rapidly due to warming temperatures and the colonial exploitation of natural resources, often tied to green economic strategies [1]. Additionally, Indigenous and non-Indigenous youth face distinct cultural and educational challenges, which further complicate their responses to climate change. These issues emphasize the importance of including diverse voices from different Arctic communities in climate-related research and initiatives [2, 3].

The antler workshop is part of the more extensive research project titled "On the Frontline of the Climate Crisis" (the climate crisis project), 2023–2025, funded by the Kone Foundation in Finland and coordinated by the Youth Research and Development Center, Juvenia, with collaborative support from the Degree Program on Art Education at the University of Lapland. The climate crisis project examines young people's views, involvement, and emotional responses to climate change, which poses significant risks to their future living conditions. Within the project, young people from the Barents region (northernmost parts of Norway, Sweden, Finland, and Northwest Russia) and the middle of Finland have been involved in art-based events and processes on a participatory research basis, thereby making the most of the opportunity to influence and have their voices heard [4].

DOI: 10.4324/9781003545880-9

We, the authors of this chapter, work as artists, art educators, and researchers in the climate crisis project. By employing an art-based action research (ABAR) approach, our research fosters empowerment and enables young people to actively contribute to discussions on climate change and propose sustainable solutions. Apart from the environmental challenges the youth encounter, this research highlights the unique possibilities art and art education have when working to create a hopeful future.

The participants in the antler workshop are Sámi youth from Sápmi (the Sámi homeland) and young people from the Barents region. The Sámi people and culture have been subject to national assimilation politics and settler colonialism for centuries. The young Sámi generation is growing up in an increasingly intruding Western culture, and their responsibility to safeguard the future of the Sámi culture is a distinct one. The Sámi youth are also more frequently having to settle in the South, away from Sápmi, due to studies and jobs, but their ties to Arctic Nature remain usually strong. Sámi culture and languages are related to nature and the Arctic environment, and Sámi people have made revitalization efforts for the endangered Sámi languages and culture since the late twentieth century [5].

The basic idea for the workshop follows the ideology of new genre Arctic art education [6, 7, 4041] realized by the theory of community-based art education [8] and the ABAR method [9]. It emphasizes new materialism and place-specific cultural knowledge rooted in Arctic art.

Our chapter frames the theory of community-based art education, in which knowledge is practical but first emancipatory, guiding toward awareness. Performativity and dialogue are central to creating communality [8]. The principle of performativity in community-based art education is linked to the transformative learning theory. The core idea of transformative learning is testing meaning perspectives to become more open, porous, receptive, and multidimensional to new insights [10]. This theory emphasizes the critical dimension of learning, which enables recognizing and reassessing the structures of assumptions and expectations that shape thought, emotions, and actions [11, 12]. In the face of changes brought about by the climate crisis, such a critical dimension is paramount.

From the perspective of community-based art education, identifying the critical dimension and the structures of assumptions and expectations is a performative event that connects with the intra-action between body and material, moments of repetition with difference, when the act of "doing" transforms into conscious action [8, p. 257]. In our chapter, we also describe the significant relationship between intra-action [13, 14] and performativity. In addition to the young participants in the workshops, our study brings forth themateriality of arts by using reindeer antlers. The intra-action between human and non-human and non-living nature creates a dialogic setting. Moreover, the importance of agency in the non-human aspects of the art event is essential in the antler workshop.

Our chapter seeks to answer the following three key questions: (A) What hopes, concerns, and meanings did the participating youth express through their art in the antler workshops? (B) What roles do materials, place-specificness, and artistic methods play in supporting actions for the Arctic's future? (C) In what ways can new genre Arctic art education empower and encourage youth participation in urgent environmental issues?

At the core of our research is an exploration of young people's perspectives on the climate crisis, examining their attitudes, emotional responses, and levels of engagement. Through our study, we expect to generate new insights into the emotional and sociopolitical dimensions of youth climate participation, which could even contribute to shaping more effective climate policies in the future. In the following section, we introduce the theoretical framework and research method, and then we discuss the nomadic course of the antler workshop in the North.

6.2 Background

Art education in the Arctic, historically tied to colonialism [15], must adapt to social and environmental changes, including climate change [1]. The relationship between Arctic sustainability research and Arctic art emphasizes the need for culturally relevant education that suits Northern contexts [16]. Education should raise awareness, challenge stereotypes, and reassess traditional art education. In this context, the key factors include land-based pedagogy and tangible experiences for Arctic children and youth [17]. Scholars from the University of Lapland have introduced the concept of new genre Arctic Art Education (AAE), which promotes activism, resilience, and cultural preservation, thus enabling communities to adapt to social and environmental changes [18]. This concept focuses on ecocultural revitalization and strengthening Northern cultural identities [7, 19]. AAE aims to empower Arctic youth through art, fostering hope and regional resilience [20]. The AAE approach views participants as co-artists and treats the art-making process as a shared learning and development project, which is in line with the principles of democratic co-research [21, 22]. The art-based activities in our study refer to social interactions that focus on creating art. These activities emphasize collaboration and shared experiences, thereby serving as platforms for addressing social, political, and environmental issues.

In the Arctic, artists often use art to oppose the exploitation of natural resources, support Indigenous rights, and address local concerns [23]. Jokela and Huhmarniemi [7] and Jokela et al. [6] define new genre Arctic art education as a form of art-based action grounded in the Northern and Arctic regions and addressing global challenges such as climate change. New genre Arctic art education fosters and produces Arctic art that is culture-specific and connected to place, time, environment, and people. It holds the potential for educational shifts toward resilience, decolonization, and sustainability [6, 7].

According to Jokela et al. [24], Arctic art includes creativity, artifacts, and handicrafts. Contemporary Arctic art is made in the Arctic by people with a cultural connection to the Arctic. While it has a solid connection to Indigenous art, crafts, and cultures, it is also open to other cultures. Arctic art mainly defines art and design practices that reflect and reform the cultural heritage of the Arctic regions. Jokela et al. [24] suggest that Arctic art has four purposes.

First, it can renew and correct false or twisted representations of Arctic cultures and environments. Second, it represents all cultures and heritage of people living in the Arctic, regardless of cultural identity and background. The Arctic is a multicultural reality that may be unknown in other parts of the world. Third, Arctic art plays a holistic role, where design, craft, art, heritage, culture, and daily matters blend and form a unique entity. Fourth, the Arctic can be considered an arena for innovative art and design development laboratories that may generate more human capital by investing in Arctic inhabitants [24].

Arctic art is often politically loaded. It addresses issues important for the artist or the community, usually tied to culture or colonial experiences. Typical for Arctic art is the firm connection to nature and the use of materials in hand. The art materials themselves tell the visual story. Indigenous peoples have been all over the Arctic since time immemorial using reindeer antlers for carving items, knife shafts, utilities, art, and artifacts [24, 25]. The antlers symbolize, associate, and connect the participants to Northern nature.

Art is a powerful source of influence and activism. Activism in art, also called artivism [26, 27], describes artworks in which the artist presents symbols, visions, sounds, or movements that strike a chord with the viewers' emotions, thoughts, and perspectives. Williams [26] describes artivism as unconventional artistic forms that question social conventions, change, and prejudices and raise critical public discourse on the status quo. Scholars suggest artivism as an alternative to more conventional forms of activism, using visual performance and arts capacity to arouse emotions and unite communities. Artivism provides a platform on which artists and activists work together to create a more just world [26, 27].

The Indigenous worldview and conception of art are holistic, and they differ from the Western understanding of art. People of the First Nation have developed land-based education to maintain and revitalize Indigenous cultures [28, 29]. Wildcat et al. [28] emphasize that the word "land" in this approach refers to the communities' land of identity and culture, composed of, for example, life memories, songs, stories, feelings, smells, and weather experiences. Land-based education is related to the relationship of this symbolic and abstract land, which is a fundamental starting point for a holistic understanding of our cultural environment. Land-based education is often operated in the environment, on the spot, with an experimental and practice-based approach. It combines learning through all senses. Students learn from older adults and the community [28]. Sámi scholars have introduced similar thoughts to enhance and revitalize Sámi culture. Indigenous people agree that education in a classroom setting offered in Indigenous languages does not encompass all necessary cultural understanding. Land-based education emphasizes hands-on experiences in real life and communication with the community. The state educational system can seldom provide such a holistic approach [30].

New genre Arctic art education can offer educational practices similar to land-based education, reform cultural heritage, and empower and revitalize Indigenous cultures by focusing on art-based actions. In Sápmi, the traditional

Sámi handicraft "duodji" is a fundamental source of passing on culture. Duodji is a Davvi Sámi (Northern Sámi language) word for artistic handicraft and material knowledge. Duodji does not only refer to the crafted item but also to the individual skill, Indigenous Sámi knowledge, and the heritage of the artist. Duodji and the Sámi languages are essential for the culture and by revitalizing these practices, it can be possible to keep the Sámi culture solid and vivid. In the past, every Sámi person usually had skills in duodji [25]. Today, interested Sámis specialize in different skills and sectors of duodji. For example, Duojár (Sámi artist) Petteri Laiti has developed traditional Sámi duodji into a distinct and exclusive handicraft that represents contemporary Sámi people [31]; the sculptor Per Isak Juuso—who works with wood, bone, metal, and textiles—is known for developing duodji skills and transforming traditional items into metaphors [32].

Our antler workshop is an example of new genre Arctic art education, using duodji material to promote Indigenous values and rights in the times of the climate crisis. New genre Arctic art education values ancient Arctic ecocultural knowledge related to place-specific materials, history, and cultural features. In addition, it values community wisdom and art-based activities that are planned and designed according to the local people's skills, knowledge, and place-specific history.

6.3 Method and Research Design

A core element of our project is using artistic interventions to foster youth empowerment. The research highlights a novel approach to promote climate action and sustainability by involving young people in art creation processes. Art-based research methods [33, 34] can collect tacit experience information not conveyed by traditional research methods based on spoken or written language. The aim of artistic activities here is to offer an avenue for expressing personal and collective visions of the future, strengthening young participants' sociopolitical agency. Our research aims to develop ways to facilitate and promote a youth-centered perspective on climate change. The ABAR strategy has been developed alongside art-based activities to enhance the development of analysis and sharing of results [9, 35].

Our project employed the ABAR approach to facilitate articulating participants' experiences and perspectives. According to Jokela [9], ABAR is a methodological approach that integrates artistic expression and creative methods into the qualitative research process. ABAR combines action research principles with various artistic forms to engage participants meaningfully, encouraging them to express their thoughts and experiences through art. Moreover, ABAR draws insights from multiple disciplines, such as environmental science and anthropology, thereby creating a holistic approach to complex challenges, particularly for Arctic youth. This approach serves as an advocacy tool that raises awareness regarding environmental concerns, Indigenous rights, and broader community issues, aiming to influence policy and promote sustainable practices [9]. ABAR has become an essential methodology for conducting research as well as analyzing and disseminating the results of these initiatives, thus ensuring that art catalyzes positive social change [15, 35].

Further, ABAR operates through iterative cycles that include goal setting, art-based interventions, reflections, data analysis, and the presentation of results through research publications and artistic productions. This approach enables researchers to collect diverse research materials, including visual artifacts, documented performances, narratives, and creative expressions. Combining these artistic expressions with reflective analysis enhances understanding and engagement throughout the research process.

The ABAR processes begin by planning art events where participants can express their thoughts through art-based actions. In our antler workshop, we operate through four iterative cycles in different locations (Figure 6.1). We collected data from the workshops and art events by documenting processes and

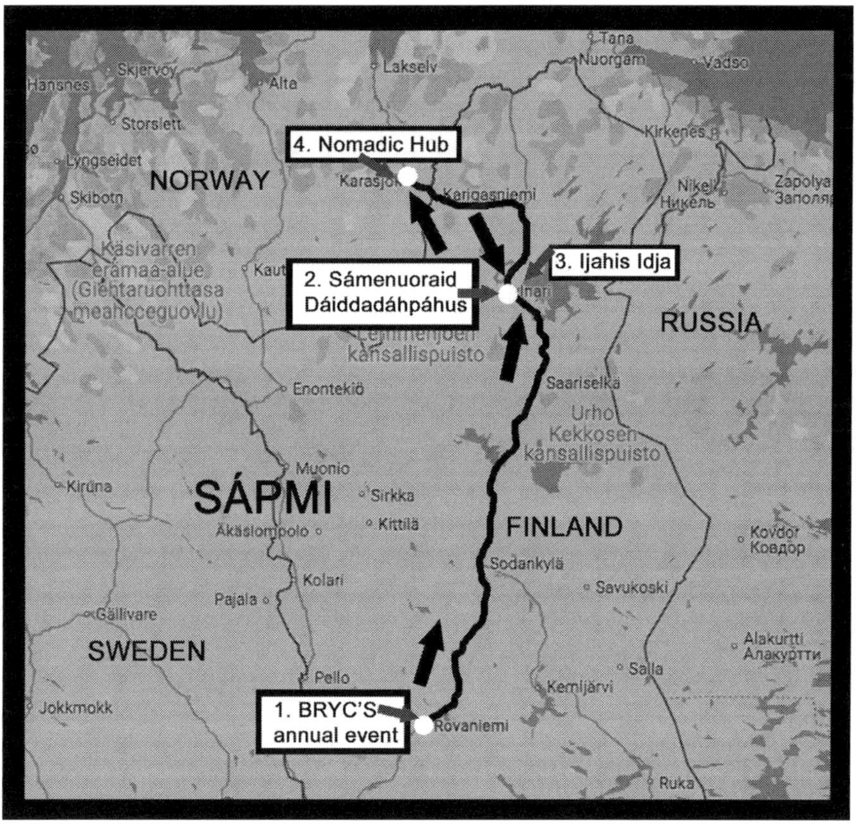

Figure 6.1 The antler workshop has been reproduced on several occasions. Picture: Korinna Korsström-Magga, 2024. The map is modified from a screenshot of Google maps (January 7, 2025).

outcomes, organizing and recording group discussions, collecting written feedback through surveys, and maintaining a research diary.

The strategy highlights place-specificness in which the local culture and material have a specific role. As is typical for ABAR, each action is a research cycle that includes reflection on experiences and, based upon these, plans adjustments for the following action or art-based event. Collecting research material requires a culturally and socially sensitive approach, and the analysis must be reflective. The reflective process aims to support the participants and their communities and develop the art educator's and researcher's skills.

The antler workshop traveled nomadically in the North. Its purpose and methods were introduced to youths of different ages and backgrounds. First, we offered the Antler workshop at the Barents youth's annual event in Rovaniemi, Finland, in February 2023. Second, in April, the antlers wandered North to Inari, Finland, where the Sámi youth participated in the workshop in the *Sámenuoraid Dáiddadáhpáhus* (the Young Sámi Art event). After the summer of 2023, a workshop with the Antlers Lávvu installation was open for young schoolchildren during the *Ijahis Idja* (the Nightless Night) music festival. From here, in April 2024, the workshop wandered to the *Nomadic Hub* seminar in Karasjok, in which Sámi youth and the seminar participants continued to contribute their thoughts and wishes regarding the climate crisis and the future Arctic on the antlers. Finally, the installation found its permanent spot in Inari in front of the Sámi Parliament building, Sajos.

Part of the ABAR strategy is the planning undertaken by the art educator–researcher for the event's workshop, considering numerous factors. When it is possible to plan art-based actions on a long-term basis, the art educator can plan the workshop theme and activities together with the community and seek suitable approaches, materials, and ways of artistic expression. However, art workshops are often planned for short-term activities as part of a bigger event. The art educator–researcher envisions and plans the big picture of the art workshop and tailors the entity to fit the ambitions of the event's main theme. It is essential to highlight the place and local culture, to answer the participant's interests, to be suitable for their age and skills, and to provide room for the participant's personal expression and creativity. A well-planned art-based activity contributes to rich and diverse research material. A fundamental aim of planning our workshops was to create a space in which the participants could feel safe to act, think, discuss, and create without restraint and concentrate on their subjective endeavors within the boundaries of the workshop topic—the climate crisis and the sustainable future of the North.

6.3.1 First Cycle: The Antlers Begin Their Nomadic Route

The art-based action with antlers began at the Barents youth annual event, 2023, held in Rovaniemi at the University of Lapland. The Barents Regional Youth Council (BRYC) organizes the event every year. The events, open to all youths in the Barents region, aim to strengthen cross-border people-to-people cooperation among the Northern youth and empower young people to actively participate in their societies and ways to get their voices heard. During the annual event, the participants work on and discuss urgent topics, such as culture, sport, health, education, environment, entrepreneurship, human rights, equality, community development, and concerns of Indigenous peoples [36]. In 2023, the event's topics were green transition, the sustainable future of the Arctic, and the climate crisis. Twenty-two young participants from the Barents region, as well as 16 art education students from the Faculty of Art and Design, participated.

The participants in the event were aged 15 to 30 years old. Their interest in participating in the annual event was political and societal, and only a few came with experience of art workshops with such variation as represented in the Think Green event. Their interest in discussing the event topics—sustainability, green transition, climate crisis, and political inefficiency—with each other was notable, but not all of them were familiar with the Sámi culture, the contemporary Arctic every day, and the Northern traditional livelihoods.

The second author of this chapter, Korinna Korsström-Magga, conducted the antler workshop. She was the designer and instructor of the workshop and brought the antlers from her home in a Sámi reindeer herders' village, Kuttura. She has shared the everyday life of Sámi reindeer herders, along with her husband, for over 30 years. Over the years, she has witnessed severe changes in nature and reindeer pastures caused by human exploitation and climate change. Thick and icy snow covers prohibit the reindeer from natural grazing during winter seasons, thereby leading to the reindeers' starvation. The hot summers dry the swamps, and new bacteria and parasites attack the animals and cause new diseases that lead to their demise. For the Sámi reindeer herders, there are severe consequences of this for their livelihood, which is fundamental to their identity and culture.

The reindeer antlers as an art material brought an Arctic ambience into the workshop and led the participants to discuss climate change concerns from the perspectives of Sámi cultural livelihood and environment. Korsström-Magga brought a pile of antlers to choose from, as the reindeer shed their antlers on a yearly basis. For many participants, this was the first time they had held a reindeer antler in their hands. The material inspired the participants to ask and talk about Sámi reindeer herding as an Arctic livelihood and tradition. In addition, photographs by Suvi Kustula, a young reindeer herder, were displayed in the workshop to deepen the information regarding contemporary reindeer herding. Korsström-Magga also shared her own experiences about reindeer herding and informed the participants regarding the next steps in the workshop.

The participants expressed their thoughts and concerns regarding the climate crisis and their future in the North by carving words, slogans, symbols, and sentences on the antlers. A basic idea for the texts and symbols on the antlers was to bring out the expression of desires and good wishes. It was reminiscent of an action of votive offerings and the idea that it will ultimately

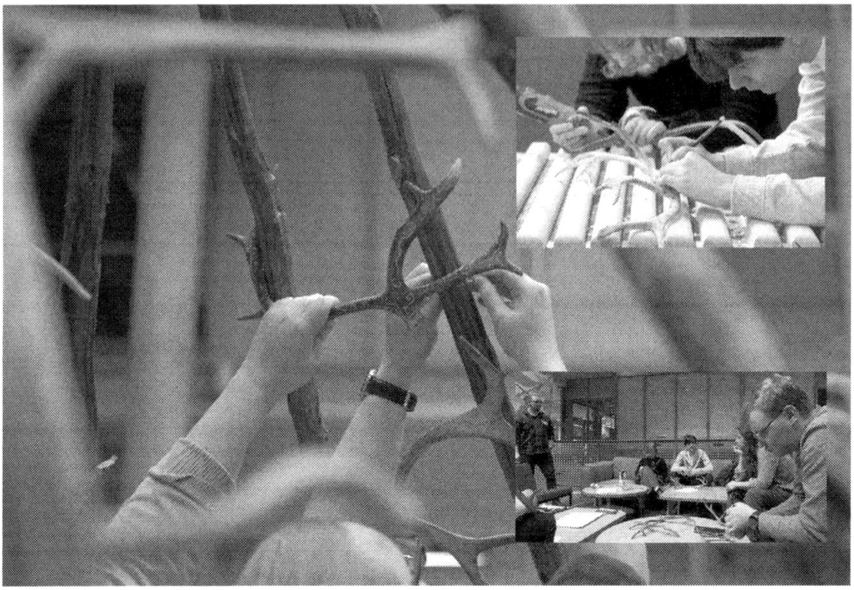

Figure 6.2 The antler workshop at the BRYC's Think Green event in 2023. Photos: Background Aki Lintumäki; upper right—Sara Poutiainen, bottom right—Taru Kervola.

come true once you have released your good wishes into the universe. After carving, they placed their antler on the mutual installation, the Antlers Lávvu. The installation was covered with antlers when the BRYC event ended (Figure 6.2).

A lávvu is a Sámi traditional dwelling, a cone-shaped tent, commonly used by the nomadic Sámi reindeer herders until the end of the nineteenth century. The symbol contributed to the Sámi people and culture, who continually endured assimilation. Today, they also give space and are giving up their traditions for the global green transition.

After the BRYC event, the Lávvu was exhibited in the University's Faculty of Art and Design hall for one month. The aim of the Antler workshop was to present the Arctic environment and raise thoughts regarding how the climate crisis and changing environment affect the Sámi culture, reindeer herding, the Northern environment, and nature.

6.3.2 *Second and Third Cycles: The Same Workshop for Different Participants*

While taking down the installation at the University, people passing by asked where the installation's permanent place would be. The question inspired us to continue the successful workshop. The idea of collecting carved antlers with thoughts and wishes widely from youth in the Arctic began to grow.

Next, the antlers wandered from Rovaniemi in the North to Inari in Sápmi, the northernmost part of Finland. The installation, as a Sámi cultural monument, found its final residence in front of the Sámi Parliament's building Sajos in Inari, Finland.

The Sámi Parliament is the political body of the Sámi in Finland. It represents the Sámi in national and international connections and is in Inari in the Sajos building. The primary task of the Parliament is to handle the Sámi people's affairs regarding their languages and culture, as well as their position as Indigenous people. In Finland, there are three different Sámi people: the Davvisápmelaš (the North Sámi), the Anarâšah (the Inari Sámi), and the Nuõrttsä′mmlaž (the Skolt Sámi). The Sajos building is also a center for Sámi cultural events and gatherings. Annually, there are over 60,000 visitors to Sajos, including local people, tourists, political decision-makers, and prominent guests [37].

The same workshop design as that in Inari and subsequently in Karasjok was introduced to the new, young participants. At Sájos, the Sámi youth participated in the workshop in the Sámenuoraid Dáiddadáhpáhus (the Young Sámi Art event) during the late spring of 2023. The event brings together all Sámi schoolchildren in Finland to present their artwork and compete with their Sámi music, acting, and dancing skills, emphasizing Sámi culture and the Sámi languages. In the event in 2023, 360 Sámi schoolchildren participated and 1064 followed the event online. It is a colorful and exciting event for all the Sámi youth, where they meet, connect, and make new friends in a face-to-face feast, sharing culture, background, and languages. During the day, they can attend art workshops apart from participating in the competition, mostly presenting Sámi traditional art and handicrafts. The schoolchildren visiting the Antler workshop were aged 12 to 15 years old, and their visits were short, as the day's program had a tight schedule, which included circulating in several places. Approximately 40 participants attended with their classmates and schoolteachers. The workshop action mainly revolved around carving the antlers with the drills. As there were only a few drills, a few schoolchildren sketched their plans or answered a short voluntary questionnaire formulated by Korsström-Magga regarding their thoughts on their Arctic future and the climate crisis while waiting for their turn to carve (Figure 6.3).

In the third cycle, the workshop with the Antlers Lávvu installation was open for young schoolchildren during the Ijahis Idja (the Nightless Night) music Festival 2023 held in the yard of the Sajos building. The Sámi Parliament's Department for Kids and Youth organized activities through the Antlers Lávvu installation, and Korsström-Magga instructed the activities. This time, the workshop sought to include minor-aged children on the day of the festival. The participants were aged 4 to 12 years old, so we left out the carving part, as we expected the drilling to be too challenging for the young participants. The children decorated the lávvu with yarn and painted antlers in traditional Sámi bright colors—red symbolizing fire, blue symbolizing water,

Figure 6.3 The youth and the children working on the Antlers Lávvu installation in the Sámenuoraid Dáiddadáhpáhus and Ijahis Idja events. Photos: Middle and left-hand side: Korinna Korsström-Magga. Right-hand side: Petra Magga-Vars, 2023.

yellow symbolizing the sun, and green symbolizing nature. During the festival evening, the lávvu was illuminated. The installation stood in the darkening evening as a festive symbol for Sámi culture and togetherness.

6.3.3 Fourth Cycle: New Genre Arctic Art Education in the Time of Climate Change

Last, the antler workshop wandered to the Nomadic Hub seminar in Karasjok, Norway, in the late spring of 2024. Here, it represented a workshop design for new genre Arctic art education. The seminar brought together art educators, researchers, and artists from the University of Greenland, Nord University in Norway, Umeå University in Sweden, and the University of Lapland in Finland to work on the concept of new genre Arctic art and its possibilities for enhancing the Arctic people's living and cultural sustainability. The seminar was held in Sámi Dáiddáguovddaš (The Sámi Center for Contemporary Art). The participants from different countries planned and introduced workshops for each other that contributed to the concept of new genre Arctic art education and discussed and reflected on its possibilities for creating a culturally sustainable future in the Arctic.

Moreover, this time, we designed the antler workshop on a pop-in basis, and the participants joined according to their time and interests. Two Sámi school classes aged 11 to 13 years old attended the seminar's art workshops. One of the workshops was the antler workshop. The pupils carved their thoughts and wishes regarding the climate crisis and the future Arctic on antlers. After the seminar, we brought the antlers carved in Karasjok to the Antlers Lávvu installation at the Sajos yard in Inari (Figure 6.4).

Figure 6.4 (1) The Antlers Lávvu installation in front of the Sajos in Inari, Finland. Photo: top left side: Korinna Korsström-Magga, 2023. Top right: Timo Jokela, 2024. (2) Each antler contributes with a unique concern, wish, or statement for a sustainable future. On the bottom left, the Antlers Lávvu exhibited in the University of Lapland. On the bottom right: BRYC participants' carvings on the antlers. Photos: Korinna Korsström-Magga, 2023.

6.4 Discussion and Results

The BRYC annual event brought together youths interested in political influence. The workshop's original design served the BRYC participants well. The workshop was appreciated, as the topic and material inspired the participants to share and discuss important matters with each other and express their thoughts by carving words and symbols on the antlers.

Through the antler workshop, the participants felt that they were able to express their thoughts directly by carving a sentence of their thoughts. After the workshop, a youth wrote, "I appreciate the combination of activism and art. It is one of my primary interests. I think art needs to be political. It is a good medium to express thoughts" (Participant, aged 20 years). The youth's frustration and helplessness in the face of global disasters were immediate, as one of the declarations carved into an antler stated, "I feel distressed and unprotected about the future."

The participants at the BRYC event were unfamiliar with using reindeer antlers as working material in an art workshop. Each antler is unique in its shape and size. Drilling the antler generated visual ideas and symbols, and additional decorations were added to the antler beyond the original sketches. A few participants were unfamiliar with using art as an expression of resistance. The workshop introduced them to a new means of influencing public opinion and decision-makers [26, 27]. Participants saw the powerful statements and activism that could emerge from the workshops. One participant said, "Such workshops should be in place—instead of radical protests [like art vandalism]—in galleries" (Participant, aged 26 years).

As a material, antlers interested the participants, and the youth unfamiliar with reindeer herding and Sámi Indigenous people asked questions and wanted to know more about Sámi culture and way of living. Korsström-Magga, who is familiar with contemporary Sámi reindeer herding, described the daily living of the people, and a few participants were Sámi and described their family traditions and experiences. The participants connected to the Indigenous culture through creative work with the reindeer antlers. The workshop achieved a similar pedagogical idea as that of land-based education and Sámi pedagogy, where community, material, and togetherness are fundamental for cultural learning [28, 29]. A participant commented, "When I was carving out ornaments on a deer antler, I contemplated on the serenity and the tranquility that Indigenous people access through close contact with nature" (Participant, aged 24 years). However, the drilling requires a steady hand and a light touch. For many, the art-based action felt liberating, letting them express their thoughts by carving words on antlers. The *Antlers Lávvu* is full of antlers carved with youth greetings, requests, and frank opinions for the audience and policymakers.

Further, during the antler workshop in the Sámenuoraid Dáiddadáhpáhus, it appeared that working with the drill was too challenging for many young teenagers and they needed assistance. For most Sámi schoolchildren, reindeer antlers were a familiar material, and reindeer herding is a common profession in Inari. Antlers have been used to make buttons and handicrafts. Reindeer herders sell antlers to Northern souvenir makers, but the carving experience and mutual installation were new ways of using antlers for the Sámi teenagers.

The Dáiddadáhpáhus was a lively event, and the young teenagers' time in the antler workshop was scheduled to be brief. The time frame and the hectic ambience did not support a conversation regarding the climate crisis, and the young participants were also too shy to express their thoughts loudly. We offered them a voluntary questionnaire to answer their thoughts on the climate crisis and the future of the Arctic. Seventeen participants completed the survey. Their written responses reflected the youth's deep concern regarding their future in the Arctic. One comment, "Don't just ponder on it. Do something!"

(Participant, aged 15 years) describes the youth's feelings about policymakers' slow or reserved actions regarding the climate crisis and the urgency of acting to protect the future Arctic. The survey also asked about their most profound concern for their future environment. Their answers covered issues regarding nature and animals, primarily the reindeer, but also the wild animals, old trees, water, and the winter season with snow.

The difference between the thoughts of the BRYC event's young participants from all over the Barents region and the Sámi schoolchildren in Inari, who live in the Arctic environment and know reindeer herding, was evident. The answers from the survey indicated that the Inari schoolchildren were greatly concerned regarding the future of their home and of the welfare of the reindeer. Their answers reveal how they are affected by all the news and information they obtain from the media, the school, and opinions in their homes regarding climate change and the exploitation of their environment. They are also concerned about the wild and untouched nature and the changing climate that affects the reindeer grazing possibilities in wintertime, as they have already experienced the losses. They also worry about the inevitable undesired change in their surrounding environment.

Antlers Lávvu was at the Ijahis Idja music festival, 2023, which is a spot for art-based actions. The minor children decorated the lávvu and the antlers in the festival activities with Sámi colors. This time the antler workshop and the Antlers Lávvu installation turned into a manifestation of Sámi culture, where the colorful yarns and painted antlers, symbolizing the Sápmi landscape, guided minor children to connect the Sámi colors, the antlers as material, and the music festival to their Sámi culture. The illuminated installation in the darkening evening was both solemn and festive.

In the fourth cycle of our research at the Nomadic Hub in Karasjok, Norway, the workshop served as an example of new genre Arctic art education. It represented the relationship between Arctic sustainability research and Arctic art—the material related to place-specific and Northern culture, which also gave substance to Indigenous land-based understanding. The adults at the Nomadic Hub seminar, who were all art educators or artists, were inspired by the material and the carvings. The Sámi schoolchildren from Karasjok, visiting the Nomadic Hub's art workshops, made short visits to the workshops, mainly testing the material and contributing to the topic of the climate crisis by sketches revealing their awareness of humankind's forthcoming misfortune. In the drawing, people wear masks because of infectious diseases and their environment; the Arctic is demolished with high-rise buildings, wind power constructions, and trash thrown on the ground. The Karasjok schoolchildren's sketches show similar concern and awareness of forthcoming changes as those expressed in the Inari schoolchildren's answers in the survey.

The Antler installation grew performatively, step by step. After each workshop, more antlers were fastened on the mutual installation in front of the Sámi Parliament, Sajos building. The building is a significant public building in Sápmi, at the center of Sámi politics. In line with the principle of performativity in community-based art education [8, pp. 256–265], the culmination of the process, the Antlers Lávvu, a public artwork, serves as a powerful awakening that will last for years, not only for its creators and the participating youth but also for a broader audience. For the Inari villagers, the antlers are a familiar shape that belongs to their culture. The features of the workshop describe the essence of the new genre Arctic art education for the participants in the Nomadic Hub seminar in Karasjok. The material of the creative work and the contributing artwork connect the participants to the land and their culture.

The significance of reindeer herding in Sámi culture is crucial, and it is obvious that the cultural ways of life in Sámi are threatened due to climate change. We have reflected on the versatile research material according to the ABAR strategy [30]. Based on our findings, new genre Arctic art education can empower and encourage youth participation in urgent environmental issues. The antler workshop provided a safe space for Arctic youth to express their concerns and make an artivist statement regarding urgent contemporary concerns regarding the climate crisis.

6.5 Conclusion

In this chapter, we explored the intersection of art and activism among Arctic youth in the face of climate change, focusing on a unique ABAR project conducted in northernmost Europe. Through community art activities, specifically the carving of reindeer antlers by young participants in the Barents and the Sámi homeland, we delved into the emotions, concerns, and hopes of Arctic youth regarding the climate crisis. We were interested in the roles that materials, place-specificness, and artistic methods play in supporting actions for the Arctic's future.

The antler workshop, conducted in different locations for young people and representing Northern cultures from a broad perspective, brought forth substantial viewpoints on the effects of climate change. For some, the antlers and the installation represented Sámi Indigenous people and the Northern and Arctic environment. For others, thoughts of home, livelihood, safety, and distress for the suffering animals and nature came up.

Further, in our project, we discovered how art exists between fiction and fact, emotion and rationality, and it is a powerful tool for understanding and expressing the complex dynamics of Arctic youths' everyday environment. Through the workshop, art became a medium for young people in the Arctic to tangibly articulate their thoughts and feelings regarding climate change. Artistic activities opened a space for encounters and discussion. The reindeer antlers, carved with messages and visions regarding climate change and sustainability, are permanently displayed in front of the Sámi Parliament. This Antlers Lávvu installation not only represents the voices of the only Indigenous people in Europe who are at risk due to climate change, but also serves as a symbol of their agency and resilience. The installation makes this experience visible and shares the messages with a broader audience.

In conclusion, concrete artistic activities permit workshop participants to discover new tools, knowledge, and motivations to address climate change. This process was both anxiety-relieving and empowering for many participants, offering a meaningful way to connect with and act upon abstract and complex environmental issues. Our study highlights the importance of art-based approaches in facilitating youth activism and understanding climate change, active citizenship, and sociopolitical participation.

New genre Arctic art education offers a powerful way to reflect on one's agency and emotions. In the words of one of the participants, "In the community art workshops, I leaned into my hopes and wishes for the future" (Participant, aged 18 years).

Acknowledgments

The authors sincerely thank all the collaborators and workshop participants. Research funding for this project was provided by the Kone Foundation in Finland and the University of Lapland.

References

[1] Nordic Council of Ministers. Megatrends. TemaNord. 2011:527. Available from: http://norden.diva-portal.org/smash/get/diva2:702166/FULLTEXT01.pdf

[2] Hickman C, Marks E, Pihkala P, Clayton S, Lewandowski RE, Mayall EE, et al. Climate anxiety in children and young people and their beliefs about government responses to climate change: A global survey. *The Lancet Planetary Health*. 2021 Dec;5(12):863–873. doi:10.1016/S2542-5196(21)00278-3

[3] Karlsdóttir A, Jungsberg L. *Nordic Arctic youth future perspectives [Internet]*. Stockholm: Nordregio; 2015. (Nordregio Working Paper). Available from: https://urn.kb.se/resolve?urn=urn:nbn:se:norden:org:diva-4950

[4] Rinne J, Hiltunen M, Jokela T, Korsström-Magga K, Lintumäki A. Co-research through art-based action enhancing young peoples' climate citizenship. In: Gorman J, Kiilakoski T, Vallado E, Pasic L, editors. *Youth knowledge book. Youth and democracy in the climate crisis. Council of Europe*. Strasbourg Cedex, France: Youth Knowledge Books; 2024. pp. 55–70.

[5] Spangen M, Salmi A-K, Aikas T, Lehtola V-P. Sami histories, colonialism, and Finland. *Arctic Anthropology*. 2015 Feb 1;52(2):22–36. doi:10.3368/aa.52.2.22

[6] Jokela T, Huhmarniemi M, Beer R, Soloviova A. *Mapping new genre Arctic art*. Arctic Yearbook. (IS) Akureyri; 2021.

[7] Jokela T, Huhmarniemi M. Arctic art education in changing nature and culture. *Education in the North*. 2022 Nov 16;29(2):4–27. doi:10.26203/55f2-1c04

[8] Hiltunen M. Yhteisöllinen taidekasvatus: Performatiivisesti pohjoisen sosiokulttuurisissa ympäristöissä. (Community-based art education through performativity in Northern sociocultural environments) [dissertation]. Rovaniemi (FIN): The University of Lapland; 2009.

[9] Jokela T. Arts-based action research in the North. *Oxford Research Encyclopedia of Education*. 2019 Aug 28. doi:10.1093/acrefore/9780190264093.013.522

[10] Anttila E. Ihmis- ja oppimiskäsitykset taideopetuksessa. Taideyliopiston Teatterikorkeakoulu, Helsinki: Teatterikorkeakoulun julkaisusarja 58, URN: NBN: fi-fe20201221101868. 2022 The human and learning conception in art education. The Theatre Academy of Uniarts Helsinki's publication series 58; (FIN) 2022.

[11] Mezirow JK. An overview of transformative learning. In: Illeris K, editor. *Contemporary learning theorists in their own words*. 2nd ed. London: Routledge; 2009. pp. 90–105.

[12] Laininen E. Transformatiivinen oppiminen ekososiaalisen sivistymisen mahdollistajana. Ammattikasvatuksen Aikakauskirja, (Transformative learning as a possibility for ecosocial education). *Journal for Vocational Education*. 2019 Jan;20(5):16–38. Available from: https://journal.fi/akakk/article/view/84515.

[13] Barad KM. *Meeting the universe halfway: Quantum physics and the entanglement of matter and meaning*. Durham, NC: Duke University Press; 2007.

[14] Barad KM. Posthumanist performativity: Toward an understanding of how matter comes to matter. *Signs: Journal of Women in Culture and Society*. 2003 Mar;28(3): 801–831. doi:10.1086/345321

[15] Jokela T, Hiltunen M. Community-based art education: Promoting revitalization and eco-cultural resilience for cultural sustainability. In Lin C-C, Sinner A, Irwin RL, editors. *Community arts education: Transversal global perspectives*. Bristol: Bristol Intellect Ltd.; 2023, pp. 86–98.

[16] Hiltunen, M., Kallio-Tavin, M., & Sohlman, A. Toward place-specific and situational Arctic visual culture learning in Finnish art teacher education. *Studies in Art Education*, 2021;62(3):266–285. doi:10.1080/00393541.2021.1936428

[17] Rautio P, Stenvall E, editors. *Social, material and political constructs of Arctic childhoods: An everyday life perspective*. Singapore: Springer Nature Singapore Pte Ltd.; 2019.

[18] Jokela T, Hiltunen M. New genre Arctic art education as a way of knowing with the North. In: Jokela T, Huhmarniemi M, Burnett K, editors, *Relate north: New genre Arctic art education beyond borders*. (PT) Viseu: InSEA Publications; 2024. pp. 12–37.

[19] Korsström-Magga, K., Jokela, T. Art-based knowing with Sámi reindeer herders: A step towards resilience. In: T. Jokela, & G. Coutts, editors, Relate North #9; 2022. pp. 151–170.

[20] Priebe J, Lempinen H, Vikström H, editors. *Arctic sustainability transformation*. (SE) Umeå: Umeå University; 2023.

[21] Rättilä T, Hokkanen M, Sillanpää O. Co-constructing knowledge of young refugees' lives in Finland: Epistemological notes. In Honkatukia P, Rättilä T, editors, *Young people as agents of sustainable society: Reclaiming the future*. New York: Routledge; 2023. pp. 178–191. (Youth, Young Adulthood and Society). doi:10.4324/9781003345114-13

[22] Tokola N, Rättilä T, Honkatukia P, Mubeen FE, Sillanpää O. Participatory research among youth – Too little, too much, too romanticized? *Young People as Agents of Sustainable Society*. 2023 Jun 7;161–176. doi:10.4324/97810033 45114-12

[23] Huhmarniemi M, Jokela T. Arctic arts with pride: Discourses on arctic arts, culture and sustainability. *Sustainability*. 2020 Jan 14;12(2):1–21. doi:10.3390/su12020604

[24] Jokela T, Huhmarniemi M, Hautala-Hirvioja T. Preface. In: Huhmarniemi M, Jokela T, Hautala-Hirvioja T, editors. *Arctic arts in the time of change*. (FIN) Rovaniemi: University of Lapland; 2019. pp. 6–12.

[25] Guttorm G. Duodji: A new step for art education. *International Journal of Art & Design Education*. 2012 May 27;31(2):180–190. doi:10.1111/j.1476-8070.2012.01712.x

[26] Williams GT. Artivism: The artists and artwork fighting for social change. Sound of Life [Internet]. 2022 Nov 19. [cited 2024 Oct 18]. Available from: https://www.soundoflife.com/blogs/experiences/artivism-artists-artwork-social-change

[27] Funderburk A. Artivism: Making a difference through art [Internet]. [cited 2024 Oct 5]. Available from: https://www.artandobject.com/articles/artivism-making-difference-through-art

[28] Wildcat M, MacDonald M, Irlbacher-Fox S, Glen C. Learning from the land: Indigenous land based pedagogy and decolonization. *Decolonization: Indigeneity, Education & Society*. 2014 Dec 1;3(3):1–15.

[29] Wilson A, Murray J, Loutitt S, Scott RN. Queering indigenous land-based education. *International Explorations in Outdoor and Environmental Education*. 2021;219–231. doi:10.1007/978-3-030-65368-2_11

[30] Keskitalo P, Määttä K, Uusiautti S. *Sámi education*. Frankfurt am Main: Peter Lang GmbH, Internationaler Verlag der Wissenschaften; 2013.

[31] Artist in Focus - Petteri Laiti [Internet]. 2022 [cited 2024 Oct 29]. Available from: https://samidaiddaguovddas.no/en/10937/

[32] Lávki D [Internet]. 2016 [cited 2024 Oct 29]. Available from: https://samidaiddaguovddas.no/en/per-isak-juuso/

[33] Leavy P. *Method meets art: Arts-based research practice*. New York City: Guilford Press; 2009.

[34] Leavy P. *Research design: Quantitative, qualitative, mixed methods, arts-based, and community-based participatory research approaches*. New York City: Guilford Press; 2017.

[35] Jokela T, Hiltunen M, Huhmarniemi M. Art-based action research: Participatory art education research for the north. In: Sinner A, Irwin RL, Adams J, editors, *Provoking the field: International perspectives on visual arts PhDs in education*. Bristol: Intellect; 2019. pp. 45–56.

[36] Barents Youth. BRYC. About us. [Internet] Barents Youth; 2024. [cited 2024 Oct 4]. Available from https://www.barentsyouth.org/about-us.268261.en.html

[37] Saamelaiskäräjien toimintakertomus 2022. [The Sámi Parliament's annual report 2022] [cited 2024 Oct 5]. Available from: Saka%20tk%202022_Saka19062023.pdf

7 Knowing with the Sea Trout

Place-Specific Artwork for Addressing Environmental Conflict

Timo Jokela

7.1 Introduction

In creating the work of *Kueri's Journey*, I simultaneously worked as an artist and researcher. I examine my working processes as an arts-based method, through which I aim to bring forth the layers and connections of northern ecoculture in a polyphonic manner within the spirit of posthumanism and new materialism. Although my experiential horizon on the subject is Northern because of my background, I also consider my work within the context of the continuum of Western art.

I have studied and worked as an artist, and many of my views are based on the tradition of contemporary art. One of the most influential figures for me is the German multidisciplinary artist Josef Beuys (1921–1986), who developed the concept of social sculpture [1]. The ecological emphasis of Beuys on art and his argumentation through installation art are among the factors that inspired me to develop communal environmental art. In addition to Beuys' social thinking, my work is also connected to his well-known works performed with animals. His performances with animals have often been interpreted as shamanistic acts referring to Indigenous cultures. With his works, Beuys disrupted the conventions related to the roles of the artist, the work, and the audience. His animal performances prompted me to reflect on my own relationship with animals as a hunter and fisherman raised in a northern village. Since the 1980s, animals have gradually become part of my art and its processes. Recently, posthumanism has brought to the forefront ways of creating art with animals [2, 3]. By placing the endangered sea trout at the center of my work, *Kueri's Journey*, I played with the idea of posthumanist knowing-with launched by Ulmer [4]: "seeing more broadly through the eye of the fish." The starting point, in addition to the important aquatic biological dimension, was that speaking about the threat of the mine through the fish and fishing might open people's mouths and perhaps their minds, especially among the local northern men.

Even more influential than Beuys on my thinking about art and art education has been Gablik's [5] critique of modern art and Lacy's [6, 7] concept of new genre public art, which elevates participation and social engagement in the practice of public art, as well as Lippard's [8] definitions of place-specific art

DOI: 10.4324/9781003545880-10

that emphasize the value of locality. The concept of Arctic art, which is central to this chapter, was introduced within the research conducted by the thematic network of Arctic Sustainable Arts and Design (ASAD) at the University of the Arctic, which I lead [9], and its role in promoting understanding between Indigenous and non-Indigenous cultures was clarified in discussions at the Arctic Arts Summit conferences [10, 11]. In this chapter, by using the concept of the new genre of Arctic art, I draw attention to activist public art and performances that engage with current Arctic issues across the circumpolar area [12, 13].

The artwork *Kueri's Journey*, which I carried out in the mountain village of Äkäslompolo, is situated in a public space along the route leading to the national park. Although the work is community and place specific and I examine it in the context of the new genre Arctic art [12, 13], its theme is also universal. I link the work to the discussion that several researchers have already opened by showing how the fears, unrest, melancholy, and stress caused by environmental crises have increased in the twenty-first century, while also considering approaches to alleviating anxiety and maintaining positive agency [14, 15]. Fundamentally, with my work and analysis, I ask what kind of arts-based activity considers the challenges of rapidly changing northern and Arctic communities, cultural fragility, and experiences of injustice and provides locals with the space to speak about their future in their own living environments, even in conflicted situations. Finally, what does this all mean for sustainability in the North and the Arctic?

7.2 Conflict with Äkäslompolo Village and the Hannukainen Iron Mine Plans with the Eyes of Sea Trout

The community art environmental project *Kueri's Journey* was created in the village of Äkäslompolo—located in the municipality of Kolari in Northern Finland—during the summer of 2018 as part of the Art Äkäslompolo event [16]. Each artwork created during the event was a piece of socially and environmentally engaged contemporary art that, in line with local concerns, critically examines the proposed Hannukainen iron ore mine near the village and Pallas-Yllästunturi National Park.

This community art event, which was led by artists, was first held in the summer of 2017. In 2018, it was hosted by the café entrepreneur and puppet artist Lea Kaulanen. Artists worked with villagers to create environmental art and installations on her picturesque pine forest property. The event was grounded in community and socially engaged public art [6, 7] and employed place-based strategies in community art education [16–18]. Through my artwork, I aimed to provide villagers with space to discuss and express their views and concerns regarding mining plans.

The Hannukainen mine project is a significant topic of discussion in Äkäslompolo, and it is not the only mining project under debate in the north. Mines have many connections to the lives of locals, and the topic has various perspectives [19]. Similä and Jokinen [20] note that opposition to mines is typical in tourist destinations because opening a mine would alter the landscape, which is vital for tourism. Regarding the Hannukainen mine, Jokinen [21] emphasized that the local population is deeply divided; the mine is opposed by local villages where tourism jobs are located and where the population is increasing. In contrast, the mine has supporters in the village center and further away, where tourism is less significant, and population trends are negative.

Although residents' concerns largely revolve around the economic future of the tourism industry, active mine opponents have focused on the impact of mining on the shared water areas between Finland and Sweden [22]. Pro Ylläs highlights that the mining issue extends beyond the borders of the Kolari municipality and even the nation. The wastewater from the Hannukainen mine is discharged through a 10-kilometer pipeline into the environmentally sensitive and unique Tornio-Muonio River. According to Pro Ylläs [22], the metal content of discharge water is lethal to fish, and the Hannukainen Mining Company has not investigated the effects of xanthates on the river.

Pro Ylläs has stated that the wild salmon that migrate to spawn in the Tornio-Muonio River produce approximately half of the salmon caught in the entire Baltic Sea, which are fished by all countries participating in Baltic salmon fishing. However, the focus is not on salmon; the planned mine is a particular threat to the extremely endangered sea trout of the Tornio-Muonio River. The Äkäsjoki River, which flows through Äkäslompolo village, is the most important spawning area for sea trout. The Äkäsjoki and Kuerjoki Rivers, which alternately flow as rapids, pools, and rushing torrents between rocks, are important spawning rivers for sea trout. Äkäsjoki and its tributary Kuerjoki produce 20 to 35 percent of all sea trout in the Tornio-Muonio River basin.

The Hannukainen mine is in the immediate vicinity of the Äkäsjoki and Kuerjoki Rivers, and the sea trout will not survive the deterioration of the winter habitat. With mining, groundwater levels fall, and it is feared that the winter flow of rivers will cease. In this case, the autumn-spawning sea trout's eggs, which are left in the riverbed for the winter, will be destroyed because the river ecosystem necessary for their development would disappear. Many—locals and concerned nonlocals—are concerned with the threat the mine poses to the sea trout.

7.3 Basis of New Genre Arctic Art

The new genre of Arctic art fundamentally embodies posthumanist endeavors as a way to strengthen the connection between humans and nature and promote ecological awareness. In this sense, art could also be understood as education for sustainability. Ongoing global environmental crises have posed significant challenges to educational theory and pedagogical approaches both globally [23] and in Finland [3, 24]. In the social sciences, new materialism and posthumanism have been recognized as critical emerging trends [25]. In educational research, scholars such as Ulmer [26] have advocated for the development of methods that promote knowing-with and researching with other-than-human entities. New materialism represents an interdisciplinary effort to change the foundational assumptions about the relationships between humans and the surrounding world [27]. Posthumanism is an emerging paradigm in art and art education research, here juxtaposed with the long-dominant sociocultural paradigm. This paradigm shift moves the focus from anthropocentrism to perspectives that transcend the human species, simultaneously challenging traditional artistic methods [2].

The definition of the new genre of Arctic art is intricately linked to discussions on sustainability. Both sustainable development and Arctic sustainability are defined in various ways and for different purposes [28–30]. Currently, the discourse on Arctic sustainability transformation is active [31], laying a new foundation for a new genre of Arctic art. In the Arctic region, the concept of sustainability is politically contentious, particularly in terms of cultural and regional policies [10]. Discussions often focus on natural resources, ecological and economic dimensions, occasionally addressing social issues, while culture frequently receives less attention. Huhmarniemi and Jokela [10] highlighted that cultural sustainability is a crucial fourth dimension of Arctic sustainability—not only from the perspective of Indigenous cultural traditions but also non-indigenous. When discussing Northern and Arctic sustainability, the multiculturalism and multiethnicity of the region must be considered.

Soini and Birkeland [32] explored definitions of cultural sustainability, arguing that it should be considered either as an independent dimension of sustainable development or as part of ecological, social, and economic approaches. In this chapter, I emphasize the place-based approach categorized by Soini and Birkeland [32], where each location involved in artistic activity must assess its own needs, rights, and responsibilities for a more sustainable life. Many researchers have recognized that in the north, local and traditional knowledge [33, 34], particularly Indigenous knowledge [35], have made significant contributions to Arctic sustainability transformation.

Within the ASAD network of the University of the Arctic, the concept of Northern knowledge has been used to describe the understanding and knowledge based on the ecoculture of the region, which consists of shared traditions, social systems, and the sustainable use of natural offerings [10, 11, 35]. This includes examining how cultural heritage [36–38] or relationships with the landscape [39] manifest and how understanding these manifestations can promote the development of northern rural and sparsely populated areas in more sustainable ways [40–42]. Northern knowledge does not inherently guarantee the sustainable use of natural offerings; rather, it forms the basis for continuous learning toward sustainable outcomes. According to Huhmarniemi and Jokela [10, 11], Northern knowledge includes tacit knowledge related to material and spiritual cultures. In arts and crafts, this often manifests as visual language and stories that can be shared with the new generation and new residents of the area. Northern knowledge differs from indigenous knowledge because local knowledge integrated into ecoculture and living traditions is also created, maintained, and transmitted by other residents of the region. In the concept of

Northern knowing, the ways in which community knowledge is produced are integrated into current scientific and artistic efforts toward developing methods of knowing-with.

In this chapter, other key theoretical concepts I use to elucidate my art intervention fulfilling the agenda of the new genre Arctic art are decolonization [12, 13] and resilience [43]. I will explore these concepts in more detail in the discussion section.

7.4 Methodology: Arts-Based Action Research

Methodologically, the *Kuerin Matka* project is characterized as arts-based action research, with methods developed at the University of Lapland to address the knowledge interests of art education, community art education, and applied arts [44–46]. Arts-based action research is more of a research strategy than a single research method. This approach aims to develop the professional practices of artist-researchers and artist-teacher-researchers. It shares common features with international artistic research, arts-based research, and action research. In all these research methods, practice and research occur simultaneously, and the research topics encompass other academic fields in addition to the arts, such as social sciences and education, or, for example, regional development work.

The strategy of arts-based action research aims to conduct research that produces practical change as well as valid and justified knowledge, understanding, and new skills necessary to promote that change. The working methods of art education and community art have been applied in the conducted studies as methods for regional development and welfare work [18]. Development studies have included place-specific and community-based art projects involving communities of villages, schools, workplaces, and businesses. Development tasks are defined together with community members, which is one of the starting points of arts-based action research. Community members are involved in the research and development process.

Jokela and Huhmarniemi [45, 46] described arts-based action research as a cyclically progressing research and development process. The first phase includes familiarizing oneself with the phenomenon being studied, defining goals and research tasks, organizing the research, planning activities, and conducting theoretical and artistic groundwork. In the second phase, the process advances to artistic work or similar interventions. In the third phase, which often overlaps with practical activities, relevant research data are collected from the activities. In the fourth cycle, the collected data are analyzed, the results are conceptualized, subsequent development goals are clarified, and follow-up actions are specified based on the achieved research results. The research data for this chapter were collected during the planning and implementation phases of *Kueri's Journey* artwork, with the primary data consisting of video recordings of the implementation and the discussions surrounding it, which also formed the basis for *Kueri's Journey* video artwork.

7.4.1 The Creation of Artwork

After being selected as one of the artists for the Art Äkäslompolo event, I began my process by choosing a theme to address the mining conflict. Instead of focusing on the conflict between tourism and mining, I aimed to highlight the local residents' eco-cultural life worlds through my artwork. I started the process by using methods typical of arts-based action research as well as visual ethnography [46, 47]. The visual survey and analysis of the village's appearance, livelihoods, traditions, and stories interested the locals and helped involve them in the research. I began by meeting with a few individuals knowledgeable about the village's history. Interaction was very natural because I came from a northern village similar to Äkäslompolo. Additionally, the *Meän Elämää* (our life) exhibition at the Yllästunturi Nature Center, which tells the story of the village from the perspectives of cultural history and natural sciences, highlights the different layers of the village's story, from life in the houses, forests, and waters, to reindeer herding to farming, all the way to the emergence of tourism. An excellent photobook about the village, *Äkäslompolo. Seitsemän tunturin kylä ajan virrassa*, by Niskakoski and Taskinen [48] also served as a guide to the themes I discussed with the residents while also shaping the artwork and building collaboration.

Based on the survey, the theme I chose combined local place names and the threat the mine posed to the waters. Kuer fell near the village rises, and the Kuer River flows alongside it. For most people, these names no longer tell their stories. "*Kueri*" or "*kujeri*" is an old local term for sea trout derived from the Sámi language, now only living in place names. The significant impact of the fish that ascends from the sea is evident from the fact that it has given its name not only to the river but also to the nearby mountain.

Kueri became an icon of the threat to water bodies in discussions, perhaps also because there had been a legal case in the past about the traditional "basketfishing" fishing method in Kuerlinkka, a waterfall, which the fishermen had won against the government. I wanted my artwork to serve as a monument to the sea trout migration in the river, which is now severely threatened. The spawning of salmonids in the cold, clean, flowing waters of the north is threatened throughout the circumpolar region, not only because of natural resource exploitation but also because of climate change. In my sketches, I combined the story of Kueri with previously documented constructions of Siberian shaman graves and bird poles, known as soul birds by the Finno-Ugric people [49]. In my sketches, Kueri became a "soul fish," which, in discussions with locals, symbolized the well-being of the river's natural state and the ascending fish populations (Figure 7.1).

I designed the implementation of the artwork to create a natural situation for locals to come and help or just to stop and stay by the working site. I chose wood and traditional log-building techniques for the artwork to engage local people in conversations while doing familiar work for them. Mirja Hiltunen documented the phases of the artwork's implementation and the discussions

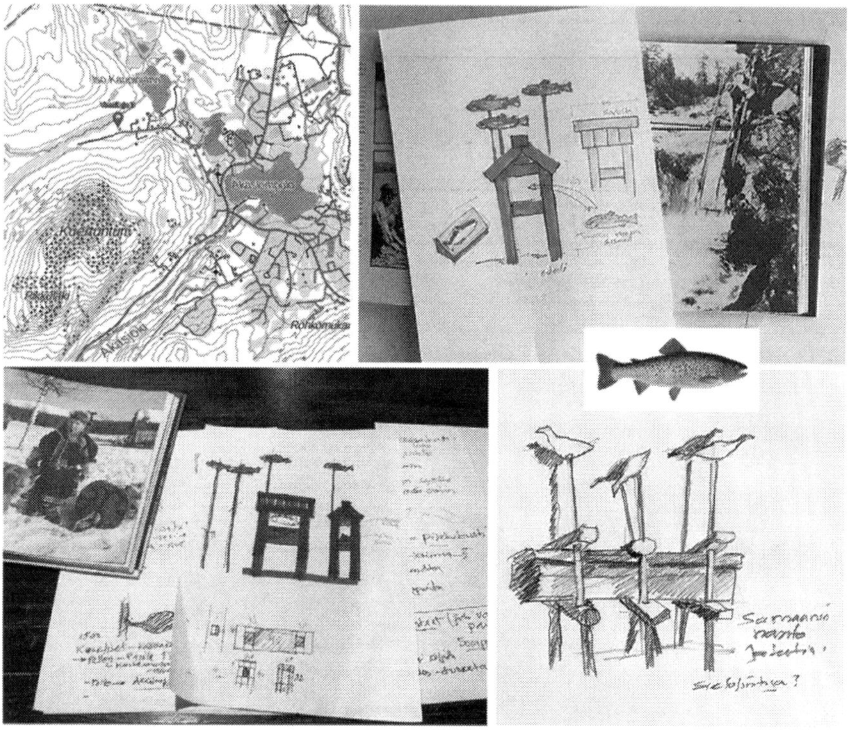

Figure 7.1 The sketches for the artwork emerge from the place, visual material, and discussions with locals. Photos: Timo Jokela, 2018; collaged by Emmi Kinnunen, 2021.

that took place during it. Conversations with locals, subtly guided and themed, ranged from the mining threat to berry picking, from fish to municipal politics, and from reflections on the eco-cultural changes in the area (Figure 7.2). The recorded conversations formed the material for both the chapter and the *Kuer's Journey* video artwork. Additionally, the artwork provided a platform to highlight the threats related to the mining plans to tourists on the national park trails, who were often unaware of them.

The story of the sea trout, or Kueri, did not end with the closing of the Art Äkäslompolo event (Figure 7.2). The artwork remains in place along the route to the national park, which is seen by thousands of people. Video artwork has been exhibited at several exhibitions, including international forums. Most importantly, local groups critical of the Hannukainen iron mine plans have eagerly continued to collaborate with the University of Lapland artists and researchers. In the fall of 2021, a willow sculpture was created by a student group invited to Äkäslompolo by mine opponents situated by the Äkäsjoki bridge in the village center. The sculpture's theme addressed the river's aquatic biology, with the sea trout once again taking center stage.

Figure 7.2 Photos collaged by Emmi Kinnunen, 2021; images from the *Kuerin Matka* video artwork (2019), scripted by Timo Jokela, filmed by Mirja Hiltunen, and edited by Ahti Nykänen. Bottom left image by Mirja Hiltunen, 2018. Audience at the Art Äkäslompolo event opening examines the finished artwork and listens to the artist's background explanation. Photo right by Timo Jokela, 2018.

7.5 Discussion around the Finished Artwork

Kueri's Journey artwork can be seen as a contemporary Beuysian social sculpture, a process created through the actions of the people involved. Emerging in a northern conflict situation, the piece also encompasses many levels of interaction between contemporary art and northern ecoculture, warranting closer examination as a new genre of Arctic art.

The roots of *Kueri's Journey* lie in environmental art, an independent art form established in the 1960s and 1970s when land and performance artists began creating artworks outside galleries and art museums. According to art scholar Malcolm Andrews [50], such environmental art pieces represent the relationship between the environment and the artist. Artwork is not just an art object placed outside; rather, its meaning arises from its relationship with its environment. Additionally, the artwork defines the character of its setting—here, the piece creates a space along the routes leading to the national park in the immediate vicinity of the Äkäsjoki River—and can be considered an environmental art piece.

Beyond the essence of environmental art, the implementation of the artwork followed the approach to the new genre of public art described by Lacy [6, 7] and the principles of place-specific community art outlined by Lippard [8]. Using these approaches, artists address social challenges specific to a community or place in their artwork and involve local actors in the artistic process. This also occurred in the realization of *Kueri's Journey*. From this perspective, my work can be considered community art, in which locals participated, as well as a public art piece situated along the trail in the national park.

A central aspect of *Kueri's Journey* was the wooden construction, which would be familiar to the men of the region. I sawed and carved the artwork over a two-to-three–day public performance, with participants and the audience joining in their own ways. In this dialogue of making and conversation, the Kueri formed a Beuysian "animal discussion partner," even though the Kueri was present only in stories and recollections. Combining craftsmanship with local fish stories provided a good starting point for natural interaction. According to Huhmarniemi and Jokela [10], a key goal of Arctic art is to find suitable ways to sensitively address environmental and sociocultural issues using the methodologies of art and art education, promoting sustainable development, and recognizing the cultural diversity of the northern and Arctic regions. Huhmarniemi and Jokela [10, 11] argued that Arctic art offers an alternative way to view art, design, and crafts as mutually supportive and integrated, rather than following the dualistic Western culture that separates art, design, and crafts into their own fields in practice, research, and education. In the design and implementation of *Kueri's Journey*, the thinking of cultural and social sustainability was present: The piece was realized in collaboration with local residents, valuing their skills, knowledge, and stories. This generated trust and openness for discussion about the mining project and its threats.

The concept of Northern knowledge, which is closely linked to the new genre of Arctic art [12, 13], refers to the interconnected ecological and cultural knowledge applied in the creation process of *Kueri's Journey*. Northern knowledge emphasizes that information is not always shared orally but can also be conveyed through visual signs, symbols, material choices, and construction techniques. This visually shaped language has communicative significance both within and outside the community. The language of craftsmanship is important in the north for sharing traditions, beliefs, and worldviews (see, e.g., [51, 52]), and local material culture can also carry spiritual and even religious content [53]. The purpose of *Kueri's Journey* was not only to convey northern knowledge but also to renew it. Northern knowledge can be seen here as a performative process, integrating it with contemporary art. Therefore, processes such as *Kueri's Journey* can also be seen as place-based regional development strategies, as described by Daniels et al. [54] and Jokela [40], that are centered around the village of Äkäslompolo in this case.

The communal implementation of *Kuerii's Journey* artwork and the appreciation of northern knowledge can also be understood as an effort toward decolonization. The execution of the work followed Smith's [38] view of decolonization as a process that, in addition to administrative measures, involves dismantling cultural and psychological colonialism. Decolonization is a step toward self-determination and community well-being. This means that locals must be involved in creating the content of cultural activities from the very beginning and be aware of the region's potential colonial history. In the context of *Kueri's Journey*, it became apparent how the residents of the mountain village of Äkäslompolo have and continue to experience colonization, from the mining company and within their own municipality. The municipality of Kolari and most of the central village residents are sympathetic to the mining project. Jokinen [55] noted that tensions exist between mountain villages and the municipality of Kolari, as well as the central village.

Decolonization is an important aspect of Indigenous cultural activism [38]. It also plays a role among non-Indigenous cultures that have lived in Lapland for centuries. Huhmarniemi and Jokela [10] suggested that similar decolonization processes pursued among Indigenous cultures should also be sought in northern communities that are non-Indigenous. Although *Kueri's Journey* contains elements of decolonization, it may be more fruitful to view it as an expression of revitalization.

Revitalization is a key effort to restore the appreciation of traditions in a prevailing changed sociocultural situation. Auclair and Fairclough [42] described revitalization as a practice that renews traditions that are part of the community's social structure. In the new genre of Arctic art, revitalization does not mean returning to a historical culture and identity that is authentic, original, or untainted. Revitalization is always based on the interpretation of history, which changes with the sources of historical knowledge as well as personal and communal perceptions, appreciations, and values. This is also the case in the village community shaped by tourism in Äkäslompolo.

A review of contemporary art in the Arctic region has revealed meaningful ways to draw inspiration from the past, such as traditional uses of materials and storytelling. It is particularly typical for contemporary artists to blur the line between art and craft and to adopt digital technology in storytelling [56–58]. In fact, contemporary art and cultural practices of our time, being products of education or cultural services, can become practices and, eventually, traditions over time. Therefore, the concept of revitalization is at the intersection of traditions and innovations and is central to the discussion of Arctic sustainability transformation [30]. In the new genre of Arctic art, this often means combining modern and traditional nature-respecting working methods [10–12]. Revitalization is also a key approach in the pursuit of cultural sustainability; it is often intergenerational and intercultural. For northern ecocultures, careful observation of the annual cycle and longer cycles, such as fluctuations in fish and animal populations, has been typical. Such observation has been the foundation for cultural practices like hunting and fishing. In *Kueri's Journey* artwork, the aim is to convey the northern knowledge system to new generations, to new community members, giving the fishing tradition a new purpose. Reminiscing did not revive the act of fishing itself but approached the strengthening of the identity of the village and the entire region. Thus, revitalization can be understood as a learning process centered on cultural interaction, which is at the core of the discussion on Arctic sustainability transformation.

In this conflicted situation, the cultural identity and ecoculture foundation of the residents of Äkäslompolo have been shaken by changes. Although mining projects are not the only force of change, tourism is likely an even stronger force. Stephen [59] saw this as a challenging situation throughout the Arctic region. After major environmental changes, residents require tolerance, adaptability, and resilience to the resulting social and cultural changes. Evaluating the process of *Kueri's Journey* artwork, it is easy to agree with Pretty's [43] view that increasing community resilience requires rebuilding the connections between culture and education and local ecocultures.

Pretty [43] suggested that our thinking be renewed by incorporating local and traditional knowledge into regional planning. Arora-Jonsson [60] also emphasized that promoting the eco-cultural resilience of communities and regions requires the recognition of local knowledge. In *Kueri's Journey* artwork, the backdrop was the northern knowledge described earlier, here based on knowing-with nature in the same way as proposed in the new materialism [25] and posthumanist research practices [4].

7.6 Conclusion

The process of *Kueri's Journey* artwork is a powerful form of place-based activity. Participatory actions revive the local ecoculture, which is becoming fragile because of changes, and the ways of knowing the landscape and waters; hence, this artwork can strengthen the identity of participants, which, in turn, impacts regional identity. Place-based art supports sustainability by reinforcing connections between people and their living environments and, as Dessein et al. [61] emphasized, by initiating discussions about communities' future aspirations. Working at the University of Lapland, the place-specific art projects I have implemented also renew higher education in the field. Responding to rapid changes in the north brings the core of the artist's and art educator's work as well as the goals of university education, from self-expression and individualism toward a more communal, dialogical, and interdisciplinary approach. The new genre of Arctic art cannot ignore the region's ecoculture, nor does it arise solely from tradition; rather, it arises from the encounter of the region's ecoculture with participatory forms of contemporary art. I believe that through processes such as *Kueri's Journey* artwork, artists and art educators can help communities transmit Arctic material and spiritual culture while renewing it to new generations, even those beyond the northern regions. Within their own ecoculture, cultural agencies act as decolonizers. Revitalization, which involves renewing traditions and stories in new contexts, also helps communities defend their rights and increases their flexibility in facing potential rapid changes. Finally, processes such as that of *Kueri's Journey* revive and emphasize northern ways of knowing. The artwork acts as a global agent in advancing nature–human connections and equality, which is the goal of global posthumanist sustainability thinking in the spirit of Arctic sustainability transformation.

References

1. Gyorody A. The medium and the message: art and politics in the work of Joseph Beuys. *The Sixties*. 2014;7(2):117–137. doi:10.1080/17541328.2015.1043800
2. Demos TJ. *Decolonizing nature: contemporary art and the politics of ecology*. Sternberg Press, Berlin; 2016.
3. Huhmarniemi M, Jokela T, Hiltunen M. Paradigm shifts in northern art, community and environment studies for art teacher education. *Social Sciences & Humanities Open*. 2021;4(1):100181. doi:10.1016/j.ssaho.2021.100181
4. Ulmer J. Posthumanism as research methodology: inquiry in the Anthropocene. *International Journal of Qualitative Studies in Education*. 2017;30(9):832–848. doi:10.1080/09518398.2017.1336806
5. Gablik S. Connective aesthetics: art after individualism. In: Lacy S, editor. *Mapping the terrain: new genre public art*. Bay Press, Seattle; 1995. pp. 74–87.
6. Lacy S. Debated territory: towards a critical language for public art. In: Lacy S, editor. *Mapping the terrain: New genre public art*. Bay Press, Seattle; 1995. pp. 171–185.
7. Lacy S. Time in place: new genre public art a decade later. In: Cameron C, Willis S, editors. *The practice of public art*. Routledge, New York; 2008. pp. 18–32.
8. Lippard L. *The lure of the local: senses of place in a multicentered society*. New Press, New York; 1997.
9. Jokela T, Coutts G. The North and the Arctic: a laboratory of art and design education for sustainability. In: Jokela T, Coutts G, editors. *Relate North: art and design education for sustainability*. Lapland University Press, Rovaniemi; 2018. pp. 98–117. http://urn.fi/URN:ISBN:978-952-310-928-5
10. Huhmarniemi M, Jokela T. Arctic arts with pride: discourses on Arctic arts, culture and sustainability. *Sustainability*. 2020;12(2):604. doi:10.3390/su12020604
11. Huhmarniemi M, Jokela T. Arctic art and material culture: northern knowledge and cultural resilience in the northernmost Europe. In: Heininen L, Exner-Pirot H, Barnes J, editors. *Arctic yearbook 2020: climate change and the Arctic: global origins, regional responsibilities?* Arctic Portal; 2020. https://arcticyearbook.com/
12. Jokela T, Hiltunen M. New genre arctic art education as a way of knowing with the north. In: Jokela T, Huhmarniemi M, Burnett K, editors. *Relate North*. InSEA publications, Viseu; 2024.
13. Jokela T, Huhmarniemi M, Beer R, Soloviova A. Mapping new genre Arctic art. Arctic Yearbook 2021; 2021.
14. Pihkala P. *Päin helvettiä? Ympäristöahdistus ja toivo* [*To hell? Eco-anxiety and hope*]. Kirjapaja, Helsinki; 2017.
15. Searle K, Gow K. Do concerns about climate change lead to distress? *International Journal of Climate Change Strategies and Management*. 2010;2:362–379.

16. Huhmarniemi M. Arts-based events for conflicted communities: engaging and educating through art. *International Journal on Education through Art*. 2021;17(2):271–280. doi:10.1386/eta_00065_3
17. Jokela T. Engaged art in the North: aims, methods, contexts. In: Jokela T, Coutts G, Huhmarniemi M, Härkönen E, editors. *Cool: applied visual arts in the North*. University of Lapland, Rovaniemi; 2013. pp. 10–21.
18. Hiltunen M. Slow activism: art in progress in the North. In: Linjakumpu A, Wallenius-Korkalo S, editors. *Progress or perish. Northern perspectives on social change*. Ashgate, Farnham; 2010. pp. 119–138.
19. Mononen T, Suopajärvi L, editors. *Kaivos suomalaisessa yhteiskunnassa* [*The mine in Finnish society*]. Lapin yliopistokustannus, Rovaniemi; 2016.
20. Similä J, Jokinen M. Governing conflicts between mining and tourism in the Arctic. *Arctic Review on Law and Politics*. 2018;9:148–173.
21. Jokinen M. Kun kaivos jakaa paikkakunnan [When a mine divides the community]. In: Mononen T, Björn I, Sairinen R, editors. *Kaivos koettuna* [*The mine experienced*]. Lapland University Press, Rovaniemi; 2018. pp. 156–160.
22. Pro Ylläs. Pro Ylläs – Ylläs ilman kaivoksia – Lapin puhtaan luonnon puolesta [Pro Ylläs – Ylläs without mines – For the clean nature of Lapland] 2021. Available from: proyllas.fi
23. Jickling B, Sterling S, editors. *Post-sustainability and environmental education*. Springers, London; 2017.
24. Värri V-M. *Kasvatus ekokriisin aikakaudella* [*Education in the era of ecological crisis*]. Vastapaino, Tampere; 2018.
25. Gamble CN, Hanan JN. What is new materialism. *Journal of the Theoretical Humanities*. 2019;24(6):111–134. doi:10.1080/0969725X.2019.1684704
26. Fox NJ, Alldred P. New materialism. In: Atkinson PA, Delamont S, Cernat A, Sakshaug JW, Williams M, editors. *SAGE research methods foundations*. Sage; 2019. doi:10.4135/9781526421036768465
27. Fondahl G, Wilson GN, editors. *Northern sustainabilities: understanding and addressing change in the Circumpolar world*. Springers, Cham, Switzerland; 2017.
28. Gad U, Jacobsen M, Strandsbjerg J. Introduction: sustainability as a political concept in the Arctic. In: Gad UP, Strandsbjerg J, editors. *The politics of sustainability in the Arctic: reconfiguring identity, space and time*. Routledge, Abingdon, Oxon; New York, NY; 2019. pp. 1–18.
29. Tennberg M, Lempinen H, Pirnes S. The resourceful North: divergent imaginaries from the European Arctic. In: Tennberg M, Lempinen H, Pirnes S, editors. *Resources, social and cultural sustainabilities in the Arctic*. Routledge, Abingdon, Oxon; New York, NY; 2019. pp. 175–181.
30. Priebe J, Lempinen H, Vikström H. *Arctic sustainability transformation 2023*. Umeå University, Umeå; 2023.
31. Soini K, Birkeland I. Exploring the scientific discourse on cultural sustainability. *Geoforum*. 2014;51(1):213–223.
32. Helander-Renvall E, Markkula I. On transfer of Sámi traditional knowledge: scientification, traditionalization, secrecy, and equality. In: Xanthaki A, Valkonen S, Heinämäki L, Nuorgam P, editors. *Indigenous peoples' cultural heritage: rights, debates, challenges*. Brill, Leiden; 2027. pp. 104–129.
33. Valkonen J, Valkonen S. On local knowledge. In: Eriksen TH, Valkonen S, Valkonen J, editors. *Knowing from the Indigenous North: Sámi approaches to history, politics and belonging*. Routledge, Leiden; 2018. pp. 12–26.
34. Virtanen PK, Seurujärvi-Kari I. Introduction: theorizing indigenous knowledge(s). *Dutkansearvvi Dieđalaš Áigečála*. 2018;3(2):1–19.
35. Jokela T. Culture sensitive participatory art as visual ethnography in the North. *Visual Ethnography*. 2018;6(2):89–112.

36. Fairclough G. Place-making and place-shaping. In: Therond D, Trigona A, editors. *Heritage and beyond*. Council of Europe Publishing, Strasbourg; 2009. pp. 29–42.
37. Fairclough G, Dragićević-Šešić M, Rogač-Mijatović L, Auclair E, Soini K. The FARO convention, a new paradigm for socially – and culturally – sustainable heritage action? *Култура/Culture*. 2014;8:9–20. Available from: https://www.academia.edu/98773699/The_Faro_Convention_a_New_Paradigm_for_Socially_and_Culturally_Sustainable_Heritage_Action
38. Smith LT. *Decolonizing methodologies: research and indigenous peoples*. Zed Books, London & New York; 1999.
39. Ingold T. The temporality of the landscape. *World Archaeology*. 1993;25(2): 152–174.
40. Jokela T, Coutts G, Beer R, Din H, Usenyuk-Kravchuk S, Huhmarniemi M. The potential of art and design to renewable economies. In: Natcher D, Koivurova T, editors. *Renewable economies in the Arctic: a state of knowledge publication*. Routledge; St. John's, Oxon; 2022.
41. Vodden K, Gibson R, Baldacchino G, editors. *Place peripheral: place-based development in rural, island, and remote regions*. ISER Books; St. John's, Oxon; 2015.
42. Auclair E, Fairclough G. Living between past and future: an introduction to heritage and cultural sustainability. In: Auclair E, Fairclough G, editors. *Theory and practice in heritage and sustainability: between past and future*. Routledge; St. John's, Oxon; 2015. pp. 1–22.
43. Pretty J. Interdisciplinary progress in approaches to address social–ecological and ecocultural systems. *Environmental Conservation*. 2011;38(2):127–139. doi:10.1017/S0376892910000937
44. Jokela T. Arts-based action research in the North. In: *Oxford research encyclopedia of education*. Oxford University Press; 2019. doi:10.1093/acrefore/9780190264093.013.522
45. Jokela T, Huhmarniemi M. Arts-based action research in the development work of arts and art education. In: Jokela T, Coutts G, Huhmarniemi M, Härkönen E, editors. *The lure of Lapland: a handbook of Arctic art and design*. University of Lapland, Rovaniemi; 2018. pp. 9–23. https://lauda.ulapland.fi/handle/10024/63653
46. Jokela T, Huhmarniemi M. Taideperustainen toimintatutkimus soveltavan taiteen kehittämisen välineenä [Arts-based action research as a tool for developing applied arts]. In: Jokela T, Huhmarniemi M, Paasovaara J, editors. *Luontokuvaus soveltavana taiteena* [*Nature photography as applied art*]. Lapin yliopisto, Rovaniemi; 2020. pp. 39–70.
47. Jokela T. Culture sensitive participatory art as visual ethnography in the North. *Visual Ethnography*.2018; 6(2); 89–112.
48. Niskakoski K, Taskinen K. *Äkäslompolo. Seitsemän tunturin kylä ajan virrassa* [*Äkäslompolo. The village of seven fells in the flow of time*]. Kirjakaari; Jyväskylä; 2021.
49. Lempiäinen P. *Pyhät ajat* [*Sacred Times*]. Kirjapaja; Jyväskylä; 2008.
50. Andrews M. *Landscape and Western art*. Oxford University Press; Jyväskylä; 1999.
51. Minnakhmetova R, Usenyuk-Kravchuk S, Konkova Y. A context-sensitive approach to the use of traditional ornament in contemporary design practice. *Synnyt/origins*. 2019;1:49–62. Available from: https://wiki.aalto.fi/pages/viewpage.action?pageId=151504259 [Accessed 26 June].
52. Schilar H, Keskitalo ECH. Ethnic boundaries and boundary-making in handicrafts: examples from northern Norway, Sweden and Finland. *Acta Borealia*. 2018;35(1):29–48. doi:10.1080/08003831.2018.1456073
53. Helander-Renval E. Animism, personhood and the nature of reality: Sami perspectives. *Polar Record*. 2009;46(1). doi:10.1017/S0032247409990040

54. Daniels J, Baldacchino G, Vodden R. Matters of place: the making of place and identity. In: Vodden K, Gibson R, Baldacchino G, editors. *Place peripheral: place-based development in rural, island, and remote regions*. ISER Books, St. John's; 2015. pp. 23–40.
55. Jokinen M. Lapin ympäristökiistojen kulttuuriset tekijät[The cultural factors of environmental conflicts in Lapland] [dissertation]. Helsinki: Helsingin yliopisto. Dissertationes Forestales; 2019. Available from: http://urn.fi/URN:ISBN:978-951-651-653-3
56. Guttorm G. Contemporary duodji: a personal experience in understanding traditions. In: Jokela T, Coutts G, editors. *Relate North: art, heritage and identity*. Lapland University Press, Rovaniemi; 2015. pp. 60–77.
57. Hansen HH. Constructing Sami national heritage: encounters between tradition and modernity in Sami art. *Konsthistorisk tidskrift*. 2016;85(3):240–255. doi:10.1080/00233609.2016.1207701
58. Igloliorte H. Hooked forever on primitive peoples: James Houston and the transformation of 'Eskimo Handicrafts' to Inuit art. In: Harney E, Phillips R, editors. *Mapping modernisms: art, indigeneity, colonialism*. Duke University Press, Durham; 2019. pp. 62–90.
59. Stephen K. Societal impacts of a rapidly changing Arctic. *Current Climate Change Reports*. 2018;4(3):223–237. doi:10.1007/s40641-018-0106-1
60. Arora-Jonsson S. Does resilience have a culture? Ecocultures and the politics of knowledge production. *Ecological Economics*. 2016;121:98–107. doi:10.1016/j.ecolecon.2015.11.020
61. Dessein J, Soini K, Fairclough G, Horlings L, editors. Culture in, for and as sustainable development. Conclusions from the COST Action IS1007 Investigating Cultural Sustainability. University of Jyväskylä; 2015.

8 Artistic Responses to Climate Change in Sakha (Yakutia)

Vera Solovyeva

8.1 Introduction

Climate change has become an increasingly important theme in public debate and the political agenda in recent years. However, many people still perceive climate change as an abstract issue that is not evident in everyday life and poses no direct threat to them [1]. To increase public awareness about the urgency of the climate crisis and to mobilize society for collective action, various visual avenues have been utilized—from the famous documentary "An Inconvenient Truth" (2006) led by former U.S. Vice President Al Gore, exhibitions in natural history museums, and scientific conferences to a variety of artistic works.

Including art in climate change communication is essential to raising awareness about this global issue and motivating people to make changes in their daily lives to reduce their carbon footprint. Art can shape how we respond to climate change [2] because it allows the audience to connect with the global process through emotions and feelings, evoking thoughts and ideas [3]. Art strengthens a sense of group identity among spectators [4, 5], mobilizes the public on behalf of science [6], and helps to overcome psychological barriers while initiating and supporting actions toward social transformation [5, 7].

Many artists, inspired by environmental activists and educators like Bill McKibben, co-founder of 350.org, became concerned about environmental justice, viewing climate change as a social problem. They engaged in the climate change genre by addressing and conveying the magnitude of the crisis [1, 8–10]. They use various art forms and interactive technologies to visualize issues caused by climate change through installations, sculptures, paintings, or multimedia projects to enhance the learning experience and encourage viewers to care about this global process. Several artists used projections of the sea rising in their paintings and installation works, such as marking the height of projected sea level rise on monuments in London with blue LED lights [1].

In Central Park, New York City, artists transformed abstract climate change predictions into an auditory and visual interface. By selecting one of four dates—1990s, 2020s, 2050s, and 2080s—a spectator initiated a sound composition that reflected the predicted temperature changes for those dates, raising its tone when temperatures rose and lowering it when they fell. The audience

DOI: 10.4324/9781003545880-11

could hear disturbing ghostly sounds, which were rising more often than falling. Central Park visitors could also see gradual changes in colors—from rusty brown to deep red, and finally to shocking pink—representing the worst climate change scenario [11]. In Boston, climate change artists created the RisingEMOTIONS installation, where they encoded survey responses of flooded communities into colors (e.g., red = angry, blue = concerned, yellow = optimistic) to share the emotions of affected communities with the public [12].

Climate change communication is essential for connecting scientists with the public because STEM subjects alone cannot tackle the problem [13]. However, there are challenges in using art for climate change communication. Artists often focus on the aesthetic value rather than the educational or ethical values of their works [14, 15]. Additionally, artists historically use analogies, narratives, and metaphoric interpretations to explain their work, but climate communication generally lacks these attributes [16]. In addition, audiences' cultural worldviews and political affiliations influence understanding, interpretation, and actions toward climate change [7]. Despite these obstacles, climate change communication has significantly risen in recent years, creating awareness among the public and engaging people in actions to address climate change impacts [17].

Unfortunately, current climate change communication narratives are becoming increasingly apocalyptic [18]. However, researchers started to doubt that negative emotions are more motivating to actions than positive emotions and argue that both emotions may trigger support for actions [5, 9, 19, 20]. Moreover, it is stressed that avoiding despair is extremely important in climate change education efforts designed for adolescents [21].

Indeed, when viewers see these stark apocalyptic images of the future while learning more about climate change issues, they may experience a range of negative emotions such as fear, sadness, anxiety, guilt, anger, and helplessness [22, 23]. The emotional response triggered by these artworks can be overwhelming and distressing, making people feel a sense of despair about the future. Some of them felt manipulated by propaganda [22]. When individuals feel manipulated or overwhelmed by the scale of the climate crisis, they can lose interest and become disengaged from taking action or advocating for changes [23, 24]. For example, there was a concern that the hacking of the University of East Anglia emails and the IPCC's false claim that all Himalayan glaciers might melt by 2035 [25] affected public opinion on climate change. The authors suggested studying the long-term effects on public trust in climate science after this unfortunate media event [26].

There are more suggestions that the main discourse of climate change communication should change and have to link to positive emotions, leading to pro-environmental behavior. More researchers argue that climate change literacy should convey that people have the power to make a difference in our future. Climate change messages need to emphasize positive progress in addressing this global problem and show best practices and ways of mitigating climate change's negative impacts [27–30]. The study provided by Yale Program on Climate Change Communication found that people who are more optimistic and hopeful that their actions can help mitigate this global process more often try and convince others to make changes and support climate change policies [19]. Some climate change artists believe that if they were able to instill in the audience the same deep love for the places and landscapes they draw, people would care about climate change because "when you love something, you want to protect it" [31].

Western climate change communication artists use scientific evidence and computer-based predictions in their art. Sakha artists engage in climate change communication from a different perspective. Indigenous peoples' ecological knowledge has developed through long-standing symbiotic (spiritual and physical) relationships with air, plants and animals, land, and mountains. This intimate connection with the surrounding landscape provides Indigenous people with detailed insight into the impacts of climate change on the environment, traditional livelihood, health, and future. Indigenous artists use dialogue between nature and people to show their personal experiences of climate change impacts, convey messages about environmental stewardship, and share their thoughts about the future.

8.2 Methods

For this case, I used semi-directive interviews with Sakha artists and personal conversations with friends and relatives. I also applied a visual research method to study Sakha art available on the internet, exploring cultural expressions, meanings, and climate change narratives expressed through the art.

8.3 Discussion

Indigenous people in the Sakha Republic (Yakutia), located in the northeastern part of the Russian Federation, are already experiencing the negative impacts of global warming. As many studies have predicted, the average temperature in the Arctic has been increasing nearly four times faster than in the rest of the world since 1979 [32]. The rising temperature has already triggered abrupt changes in the Arctic environment: permafrost is thawing, the tundra line is retreating, shifts in seasons are occurring, and others. These changes have impacted Indigenous peoples' lives in the Sakha Republic [33–42]. However, while talking to my relatives and friends in recent years and during my research in Oymyakon ulus in the Republic of Sakha (Yakutia) earlier, I noticed a rather calm attitude of Indigenous people toward climate change. Despite there being no climate change deniers in my homeland, as evidence of this global process is vivid and experienced by everyone, the source of this optimistic view is that Indigenous people believe they can adapt and mitigate the negative environmental impacts triggered by climate change, as their ancestors did. They experience negative impacts but also notice positive changes. I didn't observe any alarmist or depressing attitudes toward climate change. As many of them said: "Life is life" [38].

Environmental problems triggered by global warming are also reflected in Sakha Indigenous artists' works. However, while these artists depict dangerous changes in the environment due to climate change and express concern about the uncertain future, they usually avoid catastrophic exaggeration of events.

8.3.1 Sakha Art

In the Sakha Republic, the following types of Indigenous art have been developed since ancient times: (1) wood carving, (2) birch bark art, (3) bone carving, (4) Sakha traditional jewelry, (5) processing and sewing of fur and leather, and (6) modeling and sewing of clothes. The secrets of craftsmanship have been transmitted from generation to generation. Fine painting originated in Yakutia (Sakha Republic) in the early 20th century when local historians and ethnographers I. V. Popov and M. M. Nosov sketched the life and culture of the Sakha people. During the Soviet era, when the principles of Sovietization and industrialization were proclaimed, traditional craftsmen in Yakutsk were organized into a specialized enterprise for producing souvenirs and household items [43]. However, in rural areas, Indigenous peoples continued working with fur and practiced sewing due to the harsh environmental conditions requiring warm clothing. The design of the clothing was modernized and differed from traditional styles (from personal experience). Soviet authorities also put efforts into developing professional painting to use art for Soviet propaganda. Portraits depicted "New Era People," free from past self-consciousness, and landscapes featured development as the path to the communist future.

In 1941, the union of Sakha artists was created in Yakutia (the Sakha Republic). Then, in 1945, the professional art college was opened in the city of Yakutsk. There, painting traditions were formed under the influence of the Russian academic school [44]. The motto: "the power of art for the benefit of the development of socialism" fits perfectly for the soviet painting in that time, where the social state and authority formed the basis of almost any creative material to proclaim the construction of a bright socialist future [45]. Soviet art was filled with demands of "people and party," focusing the art on the "Socialist realism" paradigm [46].

However, traditional Indigenous art and craft making remained untouched since its meaning, design, and forms were dictated by the Sakha people's spiritual worldview and the harsh climatic conditions of the region. The Sakha people were mostly rural residents, who closely connected and interacted with nature in their everyday lives [6]. So, despite Soviet propaganda, traditional elements of culture, such as rituals and customs dedicated to household spirits, spirits of fire, home, nature, observance of fishing and hunting beliefs, etc., preserved and quietly existed through generations. Moreover, with the strengthening of the propaganda press, the interest of Sakha Indigenous artists in folklore art also increased. Traditional spirituality and worldview, livelihood activities, ancient clothing, and utensils became a source of inspiration for Sakha artists, closely tied to the culture of their ancestors [28]. Some authors called this phenomenon "cultural sovereignty" [47].

This trend only increased during and after perestroika when the Declaration on State Sovereignty was proclaimed in the Republic of Sakha (Yakutia). During that time, the concept of glasnost (open information policy) was proclaimed. This policy lifted the curtain of secrecy on political and social events in the Soviet Union [48]. After the declassification of archival materials, previously suppressed topics about historical and cultural events began to appear in numerous journalistic publications and independent TV programs [5]. Consequently, the ethnic self-awareness and ethno-cultural identity of the Sakha people grew exponentially [47, 49].

8.3.2 Climate Change Art in Sakha Republic

These days, the Sakha people still have a deep connection to nature. Many of them live in rural areas and are engaged in traditional livelihoods such as horse and cattle breeding. Even those who live in the cities maintain close ties with their relatives in villages and often visit them, helping with everyday household chores [50, 51]. As a result, all of them, including Sakha artists, are familiar with the impacts of climate change. Sakha Indigenous artists tend to express their thoughts and ideas through climate change art based on their firsthand experience of climate change impacts rather than scientific modeling. Because of this, climate change art differs radically from that in the Western world. For example, a painting by Sakha Indigenous artist Starostin Michail, "Ice Drift," from the collection "Northern Person" (Figure 8.1) shows a growing crack on the ice shelf in a place that has been intact for thousands of years. This undermines the very foundation of traditional life for Northern communities that depend on ice coverage.

This artwork reflects the reduction of Arctic Ocean ice coverage and thickness under global warming pressure [32, 52]. In July 2020, almost half of the intact ice shelf collapsed in the Canadian Arctic at the fringe of Ellesmere Island [53]. However, the message of the painting is simple and clear: yes, unexpected changes are happening, but we should not panic. We need to study problems, understand consequences, and make the right decisions. Instead of focusing on massive planetary disasters and gloomy futures, Sakha Indigenous artists tend to focus on environmental changes under climate change pressure or turn to Sakha's Indigenous worldview and spirituality. These artworks encourage viewers to look closer at the details and evoke thoughts about the consequences of people's behavior. For example, the work of the Sakha Indigenous sculptor Fyodor Markov (Figure 8.2).

The master told me the meaning of his sculpture: it represents a middle world where people live, and their homes are located (three yurts under the trunk of a world tree). The face of the white beast symbolizes nature, and its smile indicates a balance and harmony between nature and people. However, when human actions cause significant harm to the environment, such as deforestation, overexploitation of resources, pollution, and ecosystem disruption, the balance is destroyed, leading to various ecological responses, some critical to life. The beast becomes angry and destroys the yurts (on the other

Figure 8.1 (Top image) "Ice drift," by Starostin M. (photographed by Vera Solovyeva). (Bottom image) "Quinn," by Michael Starostin.

side of the sculpture). To survive, people need to find a balance between their needs and nature's needs and adopt sustainable practices.

According to Sakha legends, the loon is a shamanic bird that helped create the earth. It was sent by the god Aiyy Toyon (Lord Creator) to retrieve grains of sand and silt from the ocean floor, from which Aiyy Toyon created the earth. The sculpture (Figure 8.2, *image on right*) shows the sacred bird still holding soil in her beak as a reminder of her role in creating our earth. She desperately calls for our attention to what is happening to the land now—permafrost thaw, collapsing land surfaces, soil erosion, desertification, and other problems associated with the warming process. Due to climate change, massive wildfires occurred in the Republic of Sakha (Yakutia) in 2020 and 2021, where millions of hectares burned. The Sakha Indigenous people acknowledge wildfire as a powerful living spirit by linking the word "fire" with its dramatic actions like

Figure 8.2 (Top left) The middle world, by Fedor Markov, 2000. Picture from Fedor Markov archive. (Top right) A call of the sacred bird by Fedor Markov, 2024. Picture from Fedor Markov's archive. (Bottom left) Artist Mochkin, N., 2020. Source: https://sakhaday.ru/news/art-yakutskogo-fotografa-vstrevozhil-vlast. (Bottom right) Artist Vinokurov, A., 2020.

Source: https://sakhaday.ru/news/pytalsya-pokazat-erudirovannost-yakutskiy-twitter-predal-nasmeshkam-i-kritike-zamministra-innovaciy-yakutii

"уот сиэтэ" (the fire consumed or ate [something]) or "кытарар кыайда" (the Red won). Artwork dedicated to wildfires in the Sakha Republic is dramatic, full of tragic agony, and calls for immediate action (Figure 8.2).

Sakhaday [54] cites Mochkin: "How can we live peacefully when the world is full of indifference, selfishness, and greed? This [wildfire] represents the scale of an ecological disaster. It is hard for us to breathe." Another Sakha artist, Vinokurov, depicted a Siberian crane with flaming wings flying in the forest fires' smoke. He accompanied the art with text, stating that the fire was approaching villages, causing enormous damage to animals, Indigenous peoples, and plant life. He also stressed: "It seems to me that all residents should repost news about fires." "Maybe our government will understand that more forces and equipment are needed" [55]. During the forest fires at the beginning of 2020, people were outraged that the government did not take adequate measures to extinguish them. Indigenous people were angry about the low effectiveness of firefighting efforts, the misrepresentation of the actual scale of wildfires, the dangerous situations faced by rural Indigenous communities under threat, and the indifferent or neglectful stance of the federal government on wildfires in the Republic of Sakha (Yakutia) [39].

The figures of a horse and a Siberian crane are not chosen by artists by chance. The horse in Sakha spirituality is a cult animal with significant sacred meaning and is traditionally deeply respected [56]. In Sakha culture, the horse is not merely livestock or a companion but embodies a wild and free spirit that serves people to maintain ecological balance [57]. The Siberian crane is also a sacred bird for the Sakha people. In their traditional mythology, the Siberian crane embodies beauty and grace, expresses freedom, and serves as a mediator between the heavenly world and middle earth [58]. These artworks cry for help and immediate action to save nature.

The works of Sakha artists also highlight social problems through the relationship between nature and people under climate change pressure. For example, the painting "Quinn" by Sakha Indigenous artist Michael Starostin (Figure 8.1).

Some people, looking at this painting, could see only spring or love [43]. However, daffodils are flowers from warmer climates and are not native to Northern regions. They could grow in the Arctic region only due to global warming and globalization. The artist depicted this by showing flowers falling from the sky like rain or snow, rather than growing on the land. Look at the worried men in the shadowed background! One of them tries to hide under his crossed arms, and all of them stand in bewilderment and confusion about how to live in changing conditions. Indeed, in Northern communities, men provide food for families through traditional activities developed over thousands of years: hunting, fishing, building, and taking care of animals. Changes in land surface, seasons, plants, and animal and fish behavior due to climate change make it harder to fulfill their responsibilities. With the advancement of climate change, it may become impossible to continue their traditional livelihood activities based on Indigenous and ecological knowledge. The artwork presents this social problem without dramatic pressure, evoking thoughts on how to solve it. Overall, Indigenous art, which is based on Indigenous knowledge and worldview, offers a different angle to climate change communication art.

8.4 Conclusion

Climate change communications use art to transcend traditional academic boundaries and bridge the gap between science and emotions, enabling people to connect with the issue on a deeper level and motivating society to address the climate change crisis [1, 5, 8, 10, 59, 60]. Effective communication about climate change must be grounded in personal experience, conveyed through stories and metaphors that resonate on an individual level. As social beings, people are moved by collective norms, making it essential to frame climate action within shared values and communal responsibility. The urgency of climate change must be made clear. It is not a distant threat, but a present reality unfolding in real time. Messages should inspire action with a sense of possibility rather than loss, emphasizing solutions and gains rather than solely highlighting risks. Ultimately, motivating long-term environmental commitment requires a balance of urgency and hope, fostering a vision in which both present and future generations are empowered to act.

There is no doubt that climate change art needs to spark a dialogue that makes the average person feel they can make positive choices and thus trigger long-term changes. Effective language, storytelling, localization, visualization, and media use can help explain climate change issues, increase awareness, and enhance understanding. Four categories of effective environmental art were described: (1) a vision of a better world; (2) depicting negative scenarios; (3) using myth and folklore motives in the artworks; (4) viable outcomes and solutions after the COP21 Art Festival [27].

For Indigenous peoples, nature and culture are one and inseparable. A core aspect of Indigenous wisdom and ethics is finding balance in relationships, even amid changing conditions, and maintaining natural equilibrium. In the Western world, this way of life is called sustainable development.

Bringing Indigenous artwork into climate change communication can enrich and strengthen the message, sparking interest and dialogue from a different angle. Combining Western and Indigenous approaches in art, where the Western reflects global concerns and the Indigenous is deeply connected to local natural systems and community, may be fruitful in building a deeper understanding of the relationship between nature, people, and the consequences of our actions. This approach can better educate the public about our common responsibility in addressing climate change. Promoting Indigenous climate change art could help people connect with the issue on a deeper level, fostering an understanding of the relationship between people and nature during times of change.

References

1. Roosen, L., Klöckner, C., Swim, J. (2018). "Visual art as a way to communicate climate change: A psychological perspective on climate change–related art." *World Art* 8(1): 85–110. doi:10.1080/21500894.2017.1375002
2. Gout. E. (2021). Emotional Appeal: How Art Can Inspire Action on Climate Change, Columbia Climate School-Doherty Earth Observatory. https://lamont.columbia.edu/news/emotional-appeal-how-art-can-inspire-action-climate-change
3. Sutton, R. (2020). Art Talk with Jason Reynolds, National Endowment for the Art. https://www.arts.gov/stories/blog/2020/art-talk-jason-reynolds

4. Chen, Y., Li, S. (2009). "Group identity and social preferences." *The American Economic Review* 99(1): 431–457.
5. Klöckner CA, Sommer LK. (2021). "Visual art inspired by climate change—An analysis of audience reactions to 37 artworks presented during 21st UN climate summit in Paris." *PLoS ONE* 16(2): e0247331. doi:10.1371/journal.pone.0247331
6. Born, G., Barry, A. (2010). "ART-SCIENCE." *Journal of Cultural Economy* 3(1): 103–119. doi:10.1080/17530351003617610
7. Gruchy, J. (2020). "Christianity, art and transformation." *Acta Theologica* 29: 6–27. ISSN: 2309-9089. https://journals.ufs.ac.za/index.php/at/article/view/4937/3916
8. Nurmis, J. (2016). "Visual climate change art 2005–2015: Discourse and practice." *Wire Climate Change* 7(4): 501–5016. doi:10.1002/wcc.400
9. Sommer, LK. (2020). "'Earth' without 'Art' is just 'EH'," An environmental psychological approach on climate change communication through art, Doctoral thesis at Norwegian University of Science and Technology, pp. 299. https://ntnuopen.ntnu.no/ntnu-xmlui/bitstream/handle/11250/2684746/Laura%20Kim%20Sommer.pdf?sequence=3&isAllowed=y
10. Hahn, U., Vermeylen, F. (2023). "Mirror or hammer? News media coverage of climate-related art." *Environmental Communication* 17(2): 172–186. doi:10.1017/S0032247420000406
11. Hawkins, H., Kanngieser, A. (2017). "Artful climate change communication: Overcoming abstractions, insensibilities, and distances." Wiley Interdisciplinary Reviews: Climate Change, John Wiley & Sons 8(5), September. doi:10.1002/wcc.472
12. Aragón, C., Jasim, M., Mahyar, N. (2021). RisingEMOTIONS: Bridging Art and Technology to Visualize Public's Emotions about Climate Change in Creativity and Cognition (C&C'21). https://dl-acm-org.mutex.gmu.edu/doi/pdf/10.1145/3450741.3465259
13. Hulme, M. (2011). "Meet the humanities." *Nature Climate Change* 1: 177–179. doi:10.1038/nclimate1150
14. Bonner, J., Ramduny-Ellis, D., Peebles, D. (2012). Making Audience Experiences More Meaningful and Emotionally Engaging through Mixed Visual and Audio Media. doi:10.14236/ewic/EVA2012.20
15. Marshall, R. (2024). Thinking about Art and Aesthetic Value, Interview with Robert Stecker (the philosophy of art and aesthetic). https://www.3-16am.co.uk/articles/thinking-about-art-and-aesthetic-value
16. Moser, S. (2016). What more is there to say? Reflections on climate change communication research and practice in the second decade of the 21st century. *WIREs ClimChange* 7: 345–369.
17. Karacaoğlu, Akbaba. (2024). Multidisciplinary perspective: A review of the importance of communication in managing climate change challenges. *Environmental Research and Technology* 7(3): 457–470. doi:10.35208/ert.1461700
18. Weston, E. (2020). Climate Change and the Apocalyptic Narrative: A Critique. The Schumacher Institute. https://schumacherinstitute.org.uk/wp-content/uploads/2022/09/202009-Climate-Change-and-the-Apocalyptic-Narrative-Ella-Weston.pdf
19. Ballew, M., Marlon, J., Maibach, E., Gustafson, A., Goldberg, M., Leiserowitz, A. (2018). Americans Are More Worried about Global Warming, and Show Signs of Losing Hope. Yale University and George Mason University. New Haven, CT: Yale Program on Climate Change Communication.
20. Whiting, T. (2018). "Why we need to change the way we talk about climate change." *Medium*. https://tabitha-whiting.medium.com/why-we-need-to-change-the-way-we-talk-about-climate-change-9e43e9d77228
21. Tiron, L. (2016). «Лето снега» в день рождения, Baykal News [In Russian]. https://baikvesti.ru/new/_summer_snow__birthday

22. Gillespie, S. (2013). Climate change and psyche: Conversations with and through dreams. *International Journal of Multiple Research Approaches* 7(3): 343–354. http://mutex.gmu.edu/login?url=https://www.proquest.com/scholarly-journals/climate-change-psyche-conversations-with-through/docview/1524713352/se-2
23. Norgaard, K. M. (2009). Cognitive and Behavioral Challenges in Responding to Climate Change (May 1, 2009). World Bank Policy Research Working Paper No. 4940. Available at SSRN: https://ssrn.com/abstract=1407958
24. Salomon, E., Preston, J. L., Tannenbaum, M. B. (2017). "Climate change helplessness and the (de)moralization of individual energy behavior." *Journal of Experimental Psychology: Applied*, 23(1): 15–28. doi:10.1037/xap0000105
25. Ayar, S., Raina, V. (2022). False alarm over the Retreat of the Himalayan Glaciers, The speed and consequences of Himalayan glacier retreat have been grossly exaggerated by the media and environmental activists, CATO Institute, Policy analysis, #927. https://www.cato.org/policy-analysis/false-alarm-over-retreat-himalayan-glaciers
26. Anderegg, W., Goldsmith, G. (2014). "Public interest in climate change over the past decade and the effects of the 'climategate' media event." *Environmental Research Letters* 9: 054005. https://iopscience.iop.org/article/10.1088/1748-9326/9/5/054005/pdf
27. Smith, P. H. (2006). "Art, science, and visual culture in early modern Europe." *Isis* 97(1): 83–100. JSTOR. doi:10.1086/501102. Accessed 19 Feb 2024.
28. Petrova, A. (2002). Traditional fine art of peoples' of Yakutia, lectures, Arctic States Institute of Culture and Art, Department of Science and Higher Education of Russian Federation [In Russian]. https://agiki.ru/en/
29. Thornes, J. (2008). A Rough Guide to Environmental Art, Annual Review of Environment and Resources, Volue #33, Review article, pp. 391–411.
30. Harrington, S. (2023). 8 artists who are grappling with climate change and imagining a better world, Yale Climate Connections. https://yaleclimateconnections.org/2023/01/8-artists-who-are-grappling-with-climate-change-and-imagining-a-better-world/
31. Gregersen, T., Doran, R., Böhm, G., Tvinnereim, E., Poortinga, W. (2020). "Political orientation moderates the relationship between climate change beliefs and worry about climate change." *Frontiers in Psychology* 11: 532229. doi:10.3389/fpsyg.2020.01573
32. IPCC. (2023). Climate Change 2023: Synthesis Report. Contribution of Working Groups I, II and III to the Sixth Assessment Report of the Intergovernmental Panel on Climate Change [Core Writing Team, H. Lee and J. Romero (eds.)]. IPCC, Geneva, Switzerland, pp. 35–115. doi:10.59327/IPCC/AR6-9789291691647
33. Ksenofontov S., Backhaus N., Schaepman-Strub G. (2017). "'To fish or not to fish?': Fishing communities of Arctic Yakutia in the face of environmental change and political transformations." *Polar Record* 53(3): 289–303. doi:10.1017/S0032247417000134
34. Gorokhov, A., Fedorov, A. (2018). "Current trends in climate change in Yakutia/C." *Geography and Natural Resources* 39(2): 153–161. doi:10.1134/S1875372818020087
35. Ignat'eva, V. (2018). "Sakha Republic (Yakutia): Local projections of climate changes and adaptation problems of indigenous peoples." In: Hiyama, T., Takakura, H. (eds), *Global Warming and Human – Nature Dimension in Northern Eurasia. Global Environmental Studies*. Springer, Singapore. doi:10.1007/978-981-10-4648-3_2
36. Czerniawska, J., Chlachula, J. (2020). "Climate-change induced permafrost degradation in Yakutia, East Siberia." *Arctic* 73(4): 509–528. https://www.jstor.org/stable/26991438

37. Filippova, V. (2020). "Adaptation of the indigenous peoples to climate change effects in Yakutia: Gender aspects." *Polar Science* 26: 100596. doi:10.1016/j.polar.2020.100596
38. Solovyeva, V. "Climate change in Oymyakon: perceptions, responses and how local knowledge may inform policy." Doctoral Thesis for Environmental Science and Policy. George Mason University, 2021.
39. Vinokurova, L., Solovyeva, V., Filippova, V. (2022). When Ice Turns to Water: Forest Fires and Indigenous Settlements in the Republic of Sakha (Yakutia), Sustainability, #14, Special Issue "Long term Change and Sustainability in Arctic Social-Ecological Systems): 21. doi:10.3390/su14084759
40. Kirillina, K., Tananaev, N., Savvinova, A., Lobanov, V., Fedorova, A., Borisov, A. (2023). "Climate change impacts the state of winter roads connecting indigenous communities: Case study of Sakha (Yakutia) Republic." *Climate Services* 30: 100356. doi:10.1016/j.cliser.2023.100356
41. Popova, L., Avelova, S., Gerasimova, A., Lutz, J., Mathiesenm S., Moiakunova, A., Petrova, A., Pogodaev, M., Shadrin, V., Shishigina, A., Zhozhikov, A. (2023). Trends and Effects of Climate Change on Reindeer Husbandry in the Republic of Sakha (Yakutia), In Reindeer Husbandry, Adaptation to the Changing Arctic, Volume 1, Springer, pp. 149–186, Series Editors: James D. Ford, Priestley International Centre for Climate, University of Leeds, Leeds, West Yorkshire, UK.
42. Solovyeva, V. (2024). "Climate change in arctic and indigenous peoples: Challenges and solutions." *Ocean and Law Coastal Law Journal, University of Maine School of Law* 29(2): 317–352.
43. Tishina, T. (2004). "Folk art in Yakutia – traditions and modernity." *Nauka I Technika* #2(7): 98–105 [In Russian Language]. https://cyberleninka.ru/article/n/narodnoe-iskusstvo-yakutii-traditsii-i-sovremennost/viewer
44. Lutsenko, Y.. (2021). The modern fine arts of Yakutia: traditions and innovations, The State Museum Art Complex "The National Fine Arts Museum of the Republic of Sakha (Yakutia), Гуманитарные исследования в восточной Сибири и на Дальнем Востокеб №1. [In Russian] https://cyberleninka.ru/article/n/sovremennoe-izobrazitelnoe-iskusstvo-yakutii-traditsii-i-novatorstvo/viewer
45. Belova, K. (2019). Art or Propaganda?: What is social realism, MOST internet journal [In Russian Language]. https://mostmag.ru/iskusstvo-ili-propaganda-chto-takoe-socrealizm/
46. Unarova-Ivanova, Z. (2015). Ethnocultural traditions of Sakha Art, Anthropology & Archeology of Eurasia. Art, Identity and Ethnicity: Republics of Kalmykia, Buryatia, Altai, and Sakha (Yakutia). - London: Routledge, Taylor & Francis Group, 2015, Vol. 54, No. 3, pp. 87–98.
47. Semenova, D.. (2021). Making Space for Indigenous Perspectives: Reflections on Cultural Sovereignity in the Republic of Sakha (Yakutia). Arctic Yearbook. https://arcticyearbook.com/images/yearbook/2021/Scholarly-Papers/31_AY2021_Semenova.pdf
48. Gross, N.. (1987). "Glasnost': Roots and practice." *Problems of Communism* 36(6): 69–80.
49. Bragina, D. G. (2022). "About modern factors of Yakut identity at the turn of the XX–XXI centuries." *Etnicheskaya kultura = Ethnic Culture*, 4(3): 12–17 (in Russian). doi:10.31483/r-103550
50. Balzer, M. M. (2004). "Whose homeland is it? Shifting boundaries and multiple identities in the Russian Federation North." In E. Kasten (Ed.), *Properties of Culture: Culture as Property: Pathways to Reform in Post-Soviet Siberia*. Dietrich Reimer. https://www.researchgate.net/publication/242582402_Whose_Homeland_is_It_Shifting_Boundaries_and_Multiple_Identities_in_the_Russian_Federation_North

51. Solovyeva V, Kuklina V. (2020). "Resilience in a changing world: Indigenous sharing networks in the Republic of Sakha (Yakutia)." *Polar Record* 56: e39. doi:10.1017/S0032247420000406
52. Lindsey, R., Scott, M. (2022). Climate change: Arctic sea ice summer minimum, Climate Information and Climate Smart Nation. https://www.climate.gov/news-features/understanding-climate/climate-change-arctic-sea-ice-summer-minimum
53. Warburton, M. (2020). "Canada's last fully intact Arctic ice shelf collapsed." *Reuters*. https://www.reuters.com/article/business/environment/canadas-last-fully-intact-arctic-ice-shelf-collapses-idUSKCN2523JG
54. Sakhaday. (2020). Арт якутского художника встревожил власть. [In Russian]. https://sakhaday.ru/news/art-yakutskogo-fotografa-vstrevozhil-vlast
55. Sakhaday. (2021). "Пытался показать эрудированность." Якутский Twitter предал насмешкам и критике замминистра инноваций Якутии. [In Russian]. https://sakhaday.ru/news/pytalsya-pokazat-erudirovannost-yakutskiy-twitter-predal-nasmeshkam-i-kritike-zamministra-innovaciy-yakutii
56. Petrova, S. (2012). "The cult of the horse in the Sakha religious and ritual practice of the 19th century." *International Journal of Religion and Society* 3(3): 249–256. http://mutex.gmu.edu/login?url=https://www.proquest.com/scholarly-journals/cult-horse-sakha-religious-ritual-practice-19th/docview/1730025896/se-2
57. Duval, P. (2014). "The Yakut horse, The Long Riders Guild Academic Foundation." The world's first global hippological study. http://www.lrgaf.org/articles/yakut_horse.htm
58. Sokrut, N., Grigoryeva, I. (2023). Образ стерха в якутской культуре: проблемы экологического воспитания// Арктика XXI dек. *Гуманитарные науки*. #1(31): 74–88. [In Russian Language] https://cyberleninka.ru/article/n/the-image-of-the-siberian-crane-in-yakut-culture-ecological-education-problems/viewer
59. Jigssa, E. (2023). How Can Art Amplify Efforts to Address Climate Change? ICPAC working group. https://icpac.medium.com/how-can-art-amplify-efforts-to-address-climate-change-7d775b261f08
60. Velie, E. (2023). Can Art Change Attitudes Toward Climate Change? News/Hyperallergic. https://hyperallergic.com/827074/can-art-change-attitudes-toward-climate-change/

9 The Glutton

Brendan Griebel and Jude Griebel

9.1 Introduction

In 2022, I had the fortune of working with the Stefansson Polar archives, one of the world's largest assemblages of Arctic manuscripts located at Dartmouth College's Rauner Special Collections Library. I was there due to an interest in the collection's namesake: Vilhjalmur Stefansson, a prominent Arctic explorer who conducted multiple ambitious expeditions into the Canadian Arctic during the early 20th century. My focus was to search out and digitally return records of interest to contemporary Inuinnait communities—a regional group of Inuit in the Central Canadian Arctic for whom I work as an anthropologist—with the goal of providing source material to assist with their revival of historical language and lifeways. Stefansson was the first trained ethnographer to meet Inuinnait in 1910 and, for many of them, represented their first contact with Western culture. His writings about Inuinnait lifeways are a rare, textual glimpse into a society whose relationship with their surrounding environment would be irrevocably upset by outside ideas and technologies that Stefansson—and the influx of traders, missionaries, and government officials who followed him—imported to their world.

It seems unlikely that food should be a focus of this search for cultural knowledge, but Inuit life has always centered on the comestible. This fascination with food is not guided by culinary extravagance per se—the spectrum of traditional diet is limited, with most meals eaten raw or boiled unseasoned—but in the communion food brings with one's environment. As Inuit Elder Mary Avalak once informed me, Inuit live to eat off the land, not the other way around. "Living for food" once dictated annual migration routes, forged bonds between people and the spirit realm, and consumed the bulk of individuals' waking hours. To gather information about traditional ways of eating is to allow a better understanding of the entire ecosystem that it once anchored in place.

While I have long studied Stefansson's writing, I had not fully appreciated his penchant for documenting Inuit gustation. Throughout my time with his Polar collections, I feasted on details from fieldnotes and publications describing Inuit practicalities and preferences for eating. Observations ranged from quantities of meat consumed to the social taboos governing its consumption.

DOI: 10.4324/9781003545880-12

From these same writings, I also began to better understand where and how eating in the Arctic diverged from its traditional course. Stefansson's writing was not simply based on participant observation; he embraced Inuit food-ways with an immeasurable zeal, both incorporating them into his own Polar expeditions and promoting them to the rest of the world. From the foundations of the Inuit all-meat diet, he laid a blueprint for a future in which non-Inuit societies could expand upon Arctic traditions of consumption. He proclaimed the Arctic the future breadbasket of the globe, famously noting that "there is no northern boundary beyond which productive enterprise cannot go" [1, p. 19]. Stefansson's dream of a fruitful and prosperous Arctic defied common thinking at the time but was a contagious seed to sow. Throughout the decades, his call for Northern productivity would be echoed again and again in Canadian politics. It would be used as a platform to develop the Arctic's natural resources for the benefit of "all Canadians," but not necessarily for the benefit of Inuit.

The impact of global consumption on Inuit lives has become alarmingly clear over two decades of my living and working in the Canadian Arctic. Since 2007, I have called the territory of Nunavut home, and this has been enough time to see firsthand the impacts of climate change. Climate change affects everything: be it landscape, economy, or social health. As an anthropologist, and someone raised in a farming family, I am particularly attuned to its impact on food. Whether it is the upward tick of pricing on Northern grocery store shelves or changing strategies of hunting, the evidence of food's increasing scarcity is everywhere. Food insecurity is disproportionately worse in the North than elsewhere in the country, with roughly three-quarters of households in Nunavut citing its impact on their lives [2].

The growing scarcity of food among Inuit is not a reflection of less consumption in the Arctic, with commercial harvesting, mining and oil extraction, and other feeding frenzies continuing to be at play. While stereotyped for its barrenness, the expanse of Canada's Arctic teems with natural resources: terrestrial and marine life, fossil fuels, and minerals. The Arctic's pristine oceans abound with fish. Rare earth metals, key to the world's renewable energy revolution, are excavated *en masse* (including neodymium, praseodymium, terbium, and dysprosium), as are such minerals as lead, iron, nickel, and gold. It is estimated that roughly a quarter of the world's unexploited oil supplies reside beneath the Arctic [3]. This bounty, in a world of dwindling resources, lures the hungry masses North. According to the World Economic Forum, investments in mining, oil, natural gas, and related infrastructure in the Arctic will reach one trillion USD by 2030 [4].

The rapid onset of climate change has further enhanced the potential of Arctic offerings. Decreasing ice allows for increased shipping. Thawing permafrost and retreating glaciers grant more access to the bounties that lie beneath. The world's emissions—created through relentless industry—gravitate north to eat away at coastlines and the foundations essential to environmental and human stability. Large-scale extraction of Arctic resources, together with climate change, poses direct threats to Inuit livelihood and cultural well-being and manifests in the loss of territorial rights, biodiversity, and access to land [5, 6]. But it is in the day-to-day act of eating that many Inuit feel the impact of their changing environment most acutely. Each year, for example, an estimated 75 million kg of marine products are exported from Northern Canada to other countries around the world, while approximately 80% of Nunavut's diet is now imported from the South [7, p. 145]. The math of this equation—whether gauged in financial, environmental, or societal terms—cannot be easily reconciled.

Over months of working through Stefansson's Polar Collections, I asked myself repeatedly how acts of consumption in the Arctic could have strayed so far from the measured traditions of Inuit eating so thoroughly detailed in his texts. The question of "what happened" is one that resounds in the North, and one (as attested to by this edited volume) that artists are increasingly stepping up to address. In recognition of this, I sought clarity in the convergence of Stefansson's writing with the sculptural work of my brother, Jude Griebel. Jude is a visual artist who creates detailed figurative sculptures that visualize contemporary human entanglement with the surrounding world. In his artwork, landscapes, the species we affect, and the waste we create coalesce in vivid forms that illustrate the reach of human impact on the natural environment. For years, Jude has considered the Arctic as the site of resource extraction, highlighting the presence of new, global mouths at the Polar table, and how their bottomless hunger continues to shape the North. His visual expressions highlight not only the anthropomorphic nature of consumption, but also the scale of its insatiability. Put simply, he depicts the monstrousness of Southern excess and greed being forced on the North.

While an outsider's view into the Arctic, Jude's work shares a strong tradition among Inuit artists of using the grotesque to communicate a rapidly changing world. From the origins of Inuit commercial art, sculptors and printmakers across the north have relied on visual departure from representational imagery as a way to better understand the rapid incursion of the colonial world [8]. While early artists focused on accentuating the surreal qualities of the new technologies emerging in their midst (including airplanes or architecture), newer generations have harnessed a world of monsters to illustrate the scale of change they are witnessing. Qavavau Manumie's print *Bird-Squid Transformation* (2019), for example, depicts the fusion of two known creatures into an unfamiliar chimera wreaking vengeance on two stranded hunters. Shuvinai Ashoona's work "Tentacles" (2018) scales this writhing insecurity to cosmic levels, depicting a sea creature entangled with a pair of drowning worlds while human faces look on, implicated but helpless to intervene. In deference to these artistic traditions, Jude and my chapter seeks to bridge our respective research approaches to Arctic appetite through a story about the emergence of global gluttony. While not positioning Vilhjalmur Stefansson as a conscious instigator of this crisis, his career serves as a conduit for the Inuit diet's transition from localness to an object of forces worldwide (Figure 9.1).

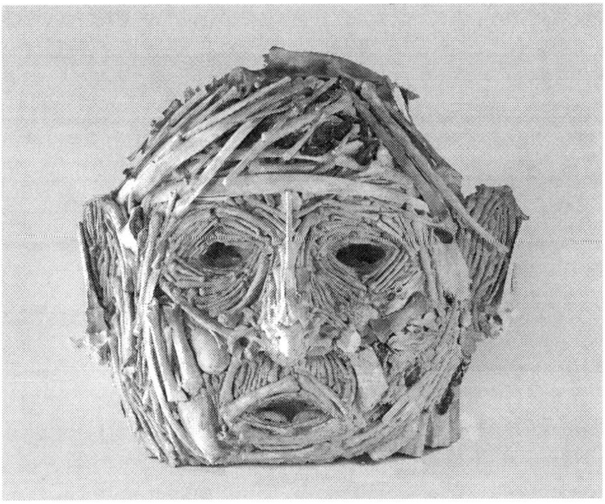

Figure 9.1 "All the Animals I've Ever Eaten" is a study of locality and consumption. It depicts an oversized sculptural head, collaged from various animal bones consumed on our family farm over the course of a year. The work functions as both a critical self-portrait and an ossuary, recognizing the many lives ingested to sustain our own being. We are literally what we eat. On a farm, there is an unspoken contract forged between animal and human. A life of care is provided in exchange for the ultimate sacrifice. There is always an awareness of quality of life: to abuse an animal, or deprive it of respect and necessities, means that one has to eventually ingest the product of one's misdeeds. This, in many ways, is not dissimilar to the relationship between hunters and the hunted in the Arctic. While acknowledging the local, this work also highlights the sense of erasure we create in Western culture when consuming other beings. Through packaging, marketing, processing, and preparation, corporeal transformations are engineered by the corporate food industry to appease our sympathies and comfort levels, as well as our palates and budgets. "All the Animals I've Ever Eaten" removes the physical and psychological barriers between the consumed and the consumer, bonding them within a single form. (Artwork by Jude Griebel courtesy of Massey Klein Gallery, New York. Photography by Brendan Griebel).

9.2 The Fat of the Land

It was in 1906, while a young anthropologist venturing into the Mackenzie Delta for the Anglo-American Polar Expedition, that Vilhjalmur Stefansson became acquainted with the Inuit diet. The ship carrying his year's supplies failed to materialize. At first, as a tentative consumer, he quickly came to understand the importance of local food for comfort and survival in the North. He documented in detail the sources and quantities of the food eaten by the Inuvialuit around him—"50 percent caribou, 30 percent fish, 10 percent seal meat, and 5 or 10 percent made up of polar bear, rabbits, birds, and eggs" by his estimate [9, p. 22]. As Stefansson's fascination with Inuit eating grew, so too did the level of detail in his descriptions. In his book "The Fat of the Land," a *magnum opus* on Inuit eating, he narrates preferences for bone marrow with an eloquence verging on sensual:

> Passing down the leg the marrow is softer and softer, more and more like a particularly delicious cream in flavor, and is in each bone softer at the lower end than at the upper, so that if one is given a small piece in the dark he can tell, by the feel when he crushes it with his tongue against his palate, and by the taste, from which bone it is and from which end of that bone [9, p. 27].

Diamond Jenness, the anthropologist on Stefansson's subsequent Canadian Arctic Expedition (1913–1918), was tasked with elaborating on the Inuit diet, paying close attention to the social dynamics of eating in the Inuit families who adopted him. He noted the intricate relationships that ensured the constant redistribution of food. Deeply rooted kinship structures obligated resources-at-hand to be shared; even unrelated men would commonly pair off as *upatitkattik*, or "flipper-associates," sending each other the hind flippers of every seal caught [10, p. 87]. Within families, food belonged to all, shared in equal measure by all members. It was also a courtesy one was expected to extend: "When a meal is in progress," notes Jenness [10, p. 90],

> every stray visitor who looks in for a moment must be offered something, even if it be only a tiny morsel of meat or fat ... if only one seal is caught, the whole of the meat must be distributed, otherwise some of the people would go hungry.

The presence of social taboos dominated the world of eating. One could be *aglirktok* (under the influence of taboo) in relation to what one ate for any variety of reasons: one's gender, age, stage of life, or name. Taboos were also distributed at the discretion of local shamans as part of their ongoing struggle to calibrate the actions of their community with the desires of the spiritual realm. The intricacy of food taboos recorded by Stefansson appears to structure all aspects of Inuit life. During his travels among the Killirmiut of Alaska, he documents the regulations guiding the cooking of caribou and mountain sheep. They must, Stefansson notes [11, p. 320], be cooked in:

> different pots on different fireplaces when cooked at same time. If cooked at successive times, the pot, if they had only one, was carefully washed with water between times. Some people never ate caribou and sheep the same meal, others ate both together. Women did not eat sheep off any of the four legs or front of the rear line of the shoulder blades. If they did, their husbands would become sick "inside" (i.e. lungs, liver, stomach, etc.). Prohibition did not apply to women past childbearing. Of the shoulder vertebrae, the women might eat the meat above the line of the

ribs, but not the fat and meat facing into the thoracic cavity below the rib line. People that ate sheep and caribou at the same meal, washed hands with water between the two courses.

The layers of consciousness and precision these taboos added to Inuit daily eating were undoubtedly overwhelming, but to disregard them was to create an open invitation for misfortune, either for oneself or the community at large (Figure 9.2).

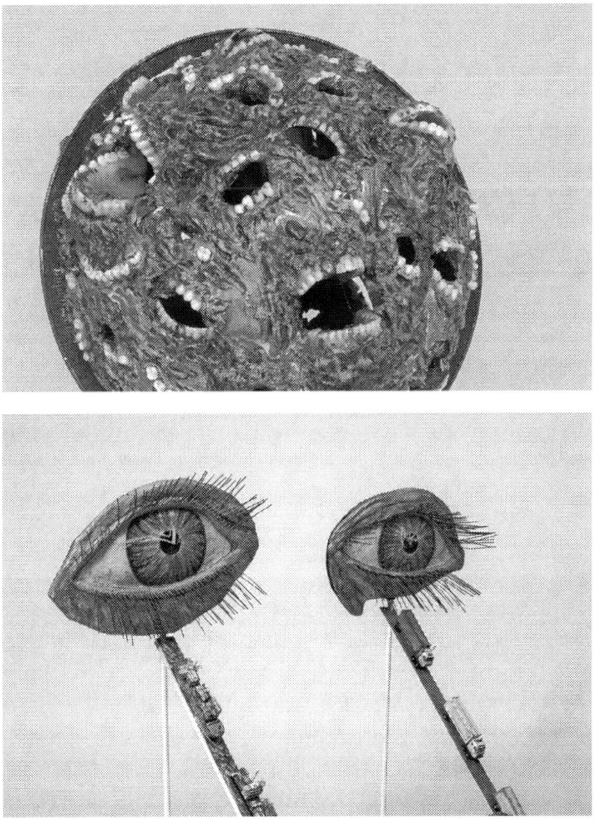

Figure 9.2 (Top image) This piece is titled "Eaten World." Scale of consumption is an underlying theme in my work. In many of my sculptures, embodied landscapes are engaged in dramatic acts of self-consumption and destruction, highlighting the fact that our world has an appetite that exceeds its ability to produce. New horizons are continually being sought to feed it. With estimates up to 40% of food produced being lost as waste along the value chain, the depletion of natural spaces for industrial agriculture and feedlots continues to increase. The work "Eaten World" depicts a terrestrial globe, the surface of which is crowded with chewing mouths, devouring both ocean and landmass. These mouths are positioned over the Arctic region, where, to some, the thawing permafrost represents untapped resources and production

(*Continued*)

Figure 9.2. (Continued)
opportunities for Southern consumption. This piece considers both the physical processes involved in eating the Arctic—resource extraction, infrastructure development, climate change—and the psychological horror of excess: bottomless appetite at the expense of people and land. My works combine scientific reality with fantasy, using personal symbolism and metaphor to register the concept of planetary collapse. (Bottom image): "Seeing and Having": In Canada, more than half of the food produced is thrown away each year, often a direct consequence of distance. This is a staggering amount, especially when considering Northern communities, where the delivery and movement of goods is increasingly becoming a necessity. In large urban centers, obtaining food often occurs via delivery, through a few clicks on a screen, rather than a physical journey to an outlet. For cultures that have been forced from nomadic ways of life, dependency on the vast distances that goods must travel from the South further underlines the shame and tragedy of colonialism. In 2024, a lonely and wilting iceberg lettuce—a distant tourist in an environment in which it could never grow—bore the price tag of $18.00. In my sculptural works, I have tried to visualize the mechanics of consumption in ways that tie us to the process of the goods we consume, from their harvesting and production to their disposal. These playful works counter the physical and psychological distance we create between ourselves and our waste, tying us back to it. In the works, tragic and sympathetic figures—often embodying landscapes—bear the brunt of our consumption habits. My 2020 work "Seeing and Having" grapples with the great distances our packaged goods travel before we thoughtlessly consume them. Ocean and highway lanes filled with cargo trucks and vessels lead from a giant set of unblinking eyes to an open mouth. Employing fantastic elements in these sculptures assists with accessibility and engagement when speaking to themes that most people do not want to pause and reflect on. (Artwork by Jude Griebel courtesy of Massey Klein Gallery, New York. Photography by On White Wall, New York).

Among the Inuit, appetite is a virtue. The presence and enjoyment of food is a celebration of scarcity's absence; overeating in times of abundance is not only a ritual, but also a right. According to the author Rachel Qitsualik [12], "there is no cap on the quantity that one can consume ... when there is enough food for everyone to eat until sated. Old, young, male, female, it doesn't matter – all descend upon the food with equal passion." This celebration of food is prompted in part by recognition that it will not always be available. "There is the lurking dread," Jenness grimly notes [10, p. 107],

> of hunger and of cold in those dark sunless days, when the huts perhaps are empty of food, the lamps extinguished for want of oil, and the people, driven indoors by the howling blizzards, huddle together on their sleeping platforms and face starvation and death.

Just as the act of eating is culturally relished, the denial of food is abhorred. The Inuit idea of gluttony, furthers Qitsualik [12], is marked not by overeating, but rather the withholding of food from others. "One of the most monstrous acts one could commit, in traditional culture, was to deprive others of food by keeping the best or largest portions for oneself." Stories of gluttons have been passed along for centuries, and through them we gain an understanding of where the limits of eating are drawn. There is the tale of Atdlarneq and The Great Glutton [13], the latter being a monstrous and insatiable being with copper cheeks. There is the parable of the miserly mother, who withholds a bounty of bear meat from her blind son so as to enjoy more of it herself [14]. Then there are stories of cannibals as the ultimate gluttons; their lust for food, as noted in another story by Rachel Qitsualik [15], "taken to such an extreme that even one's fellow humans are no longer exempt from consumption." The fearsomeness of cannibals, she notes, stems from the fact that they

> have violated this most fundamental social contract—not to prey upon those of their own species—having removed themselves to a plane of thought that is seemingly alien to the rest of us. The cannibal is a creature of chaos, choosing to remain outside of the usual social contracts that ensure human safety: in other words, a monster.

The monstrous appearance of gluttons in Inuit stories helps us understand how despicable they are perceived to be: the lumbering giant of Atdlarneq's adventure, his cheeks cast in metal for thrashing his guests. The contorted face of the blind boy's mother as she's dragged away under the water, too greedy to release the walrus harpoon line [14]. And then there are the cannibals! Rachel Qitsualik tells us of the *amayirsuk*, "a huge crone that imprisons children within the hollow hump in her back," and *nakasungnaikut*, "a man without bones in his legs, who crawls through the icy darkness hunting for normal humans, whom he ambushes and eats alive" [16]. These beings, she furthers, have appearances "so alien that there is no hope of identification with humanity."

Inuit appetite is not a one-way street. It hinges on balance and favors returned to those who fed you. Pieces of kidney and liver from each slain caribou would be dealt to the spirits as an offering, or small pieces of blubber in the case of a seal. At mealtimes, a small piece of food was sometimes "first broken off, as a sacrifice or oblation, and the remainder made the circuit of their faces before passing into their mouths" [10, p. 178]. While deeply entrenched in Inuit culture, traditions of food sharing and redistribution have been vulnerable to change. They cannot persist, Rachel Qitsualik argues, in situations where "reciprocity wanes," resulting in long-held Inuit systems becoming "singularly unworkable and un-traditional" [12].

9.3 Navigating the "Friendly Arctic"

Upon returning to New York from his last polar expedition in 1918, Stefansson's mind was still on a diet. Despite having spent more than five years not only immersed in contexts of Inuit eating but also recording their every intricacy, Stefansson's takeaway could not extend beyond the sheer materiality of consumption. The availability of resources, as opposed to the delicate relationships that ensured their continuance, became the focus of his work. Stefansson's ensuing lectures, publications, and correspondences painted the picture of a "friendly Arctic," which defied his era's preconceptions about the harshness of the North and its lack of potential to feed the masses [17]. He cited his own successes in being comfortably fed by the North as a call for

> opening new lands to the frontiersman who bring commercial development in the wake of the pioneer. "There may no longer be a Far West," noted Stefansson, "but there is a Far North with that same nebulous and glamourous future within shall rise stately cities and empires of productivity".
>
> [1, p. 67]

Citing the northward march over time of the world's great civilizations to colder and colder climates—from the Egyptians to Scandinavians—Stefansson speculated that the Canadian Arctic was the new terminus for social and economic development. "The final stage in the evolution of human history," he noted "could take place in the Canadian Arctic if Canadians were not afraid of their destiny" [17, p. 22]. This vision would be realized in great part through food. The Arctic that Stefansson spoke of was "fruitful" and, assisted through the domestication of muskox and introduced reindeer, could easily become a vast "northern prairie," ripe for commercial production [1, p. 42].

In Stefansson's vision, improved transportation of goods was essential for unlocking the Arctic's potential. When writing in 1922, he repeatedly described the Polar Ocean as a "huge Mediterranean," in terms of its potential for connecting the world through the commerce of food [1]. He predicted the future of even more advantageous technologies to enhance this network. Citing the recent successes of the Zeppelin Company, Stefansson heralded a glorious future for air travel: "We shall soon be booking our passage from New York to Liverpool by dirigible or plane or some other form of aircraft in as matter of course a way as we now book our passage by steamer (1922 p.168)." He also recognized the potential of submarines to navigate their cargo beneath the frozen expanses of sea ice, proposing the use of floating toboggans so that they could skim along the ice above them "as a fly crawls on a ceiling" (1922 p. 193). "Whoever has any grasp at all on the great natural resources of the polar regions and of the conditions under which they are about to be developed," he furthered, "will have fascinating dreams about any number of these other transportation routes destined to come into use" (1922 p. 202). It likely never occurred to Stefansson that these dreams might assume the proportions of nightmare. By the 1960s, Inuit artist Pudlo Pudlat was creating prints of cargo ships and helicopters as forces of impact on the landscape. Multiple works depict these vehicles weighed with netted caribou, muskox, and walruses, the surreal contraptions almost as animalistic as their cargo in appearance. Pudlat's foretelling vision, like Jude's sculpture "Seeing and Having," clearly outlines the dangers of transportation in its capacity for removal; the gratification of distant mouths, and denial of hungers closer to home (Figure 9.3).

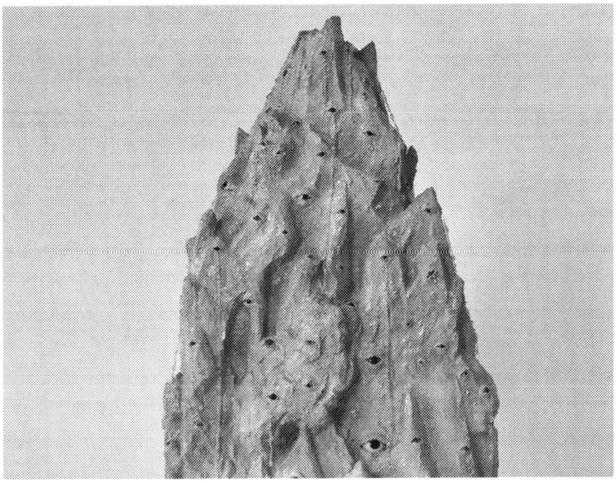

Figure 9.3 "Dissolving World": There is perhaps no greater symbol of change in the Arctic than the degradation of ice. For many Northern peoples, ice literally paves the way to survival. Sea ice, in particular, is the foundation from which one hunts, travels, and engages the natural world for a significant portion of each year. A northern world without ice is inconceivable; yet the staggering statistics to emerge from Arctic research indicate that it is inevitable. In August of 2015, a piece of Greenland's Jakobshavn glacier the size of Manhattan calved into the sea, a direct product of the ice's retreat accelerating more than threefold since the 1990s. "Dissolving World" is a monument to the impact of society's relentless hunger. A polar ice cap melts to the gallery floor, the water pooling around its base, a testament to the ongoing erosion of the natural world. The hundreds of glass eyes embedded in the ice cap's surface weep streams of transparent glass beads. A nod to Alice's dramatic pool of tears in her Adventures in Wonderland, it symbolizes a world sunken or swimming in remorse—collective culpability symbolized by the glacier's many crying eyes. This fantastic sculpture speaks to both collective grief and regret surrounding the Anthropocene and its implications for diverse human and non-human animal populations. (Artwork by Jude Griebel courtesy of Massey Klein Gallery, New York. Photography by Blaine Campbell).

9.4 Colonialism and Food Insecurity

Stefansson's vision for the Arctic's future was torn about the role of the Inuit he encountered during his expeditions. On the one hand, Stefansson greatly admired them as a model to follow, citing their soundness of health, self-reliance, and their contentment with their existing state in life. If left alone by the outside world, Stefansson openly asserted, Inuit would do perfectly alright. At the same time, however, Stefansson openly advocated for their change, speculating on what the Arctic could further become through intervention from the South.

Before Stefansson's last expedition had even left the Arctic in 1918, trading posts had moved into the area, encouraging Inuit to leave traditional subsistence hunting for the pursuit of fox furs, which could be sold to posts in exchange for western luxuries. Inuit migration routes made shorter and shorter rounds of the land, their orbit tightening around the central presence of the posts and their imported commercial goods. Key items of barter included white flour, sugar, coffee, and tea; processed and expedient foods which did little to support the physical or spiritual wellness of local populations. Other commercial technologies proffered to Inuit encouraged consumption at levels previously unachievable by their tool kits. In a matter of years, the introduction of high-power rifles had reduced the population of caribou in the Coronation Gulf—the staple food for Inuit of that region—to roughly one tenth of its numbers [18, p. 548].

By 1921, Diamond Jenness outlined the various impacts of Western society on Inuinnait since the occurrence of initial contact roughly a decade earlier, noting deep ramifications across all aspects of Inuinnait culture— diet and hunting preferences among those most dramatically altered: "Already they are becoming dependent on the outside world for tools and hunting weapons and camp equipment," he notes.

> A quarter of a century hence, it is safe to say, there will be practically no caribou at all in their country. Then, unless herds of domesticated reindeer are successfully imported and the natives trained to take care of them, either they will be forced to migrate elsewhere or, what is far more likely, they will become absolutely dependent on the outside world for clothing and to some extent also for food. Seals, no doubt, will always be plentiful, and the rivers and lakes will continue to abound with fish; but … tea and sugar and flour and rice and other cereals will become as necessary to them as meat.
>
> [18, p. 548]

Similar to Stefansson, Jenness' prophecy for an Inuit future envisioned culinary survival as contingent on the adoption of non-Inuit strategies for the production and harvesting of food. The days of living from the land, he predicted, were a thing of the past.

Over the ensuing century, the ongoing colonization of the Arctic strived to control Inuit populations through access to food. The act of eating was weaponized through the residential school system, a network of Government funded and church-delivered schools. From 1951 to 1997, a total of thirteen residential schools operated in the region of Nunavut, designed to assimilate Indigenous children through isolation from their culture. Lack of access to traditional foods was a critical part of this process. The expansion of southern interests across the Arctic swelled at the expense of Inuit mouths going unfed. As concluded by the Truth and Reconciliation Commission:

> The federal government knowingly chose not to provide schools with enough money to ensure that kitchens and dining rooms were properly equipped, that cooks were properly trained, and, most significantly, that food was purchased in sufficient quantity and quality for growing children.
>
> [19, p. 92]

So widespread and prolonged was the malnutrition of these spaces that direct connections have been drawn to a myriad ongoing health issues—from elevated risk of obesity, early-onset insulin resistance, and diabetes—that continue to plague Indigenous populations today [20].

The most current colonial threat to Arctic food sovereignty is that of climate change. While more subtle in its violence than residential schools, it is equal in its ability to deprive Inuit of the resources they require to flourish. The fruition of Stefansson's stately "empire of productivity" in the North has emerged at the expense of not only the natural environment but also the people who rely on it. The connection between an environment-in-flux and the human repercussions of hunger is still a difficult idea for many people to bridge. While the floods, hurricanes, and wildfires creep closer to home for non-Arctic

Figure 9.4 In the sculpture "All-Consuming," a giant body doubles as an island and a shipping yard. The body is being fed by an endless stream of freight vessels, arriving from different geographies to unload and deliver crates into its maw. The figure's face (referencing grotesque statuary from the Italian Gardens of Bomarzo) updates the context of the insatiable Renaissance-era Hell Mouth motif. The miniature scale of the delivery vessels, in relation to the anatomy, makes a nod to Jonathan Swift's Gulliver's Travels, in which the captive Gulliver's appetite threatens to consume everything within a miniature world. While this image presents an alternative way to envision our dependency on diverse goods and food systems in urban centers, it also speaks to environments being stripped and harvested to fuel consumption habits. A future in which the Arctic is stripped of its minerals and resources for Southern appetite calls forth images of the Hell Mouth, the Glutton, and monstrous, unending hunger represented as mythological beings in numerous cultures. (Artwork by Jude Griebel courtesy of Massey Klein Gallery, New York. Photography by On White Wall, New York).

populations, many often still have food on the table. This conceptual divide is where the role of the artist can flourish. The sense of human tragedy behind ecological disaster is clearly represented in Jude's *Dissolving World*. It is visceral in its parallel between human tears and the run-off from ice caps. A plastic bottle of "glacial water," once consumed so casually, becomes a little harder to swallow. In a similar vein, Ningiukulu Teevee's 2001 ink drawing *Untitled (Sedna by the Sea)* intercepts our thoughtless consumption of goods by highlighting the less often-seen waste it creates. In it, a tired looking Sedna—the goddess once relied on by Inuit for successful procurement of food from the sea—perches ocean-side smoking, while teams of dump trucks empty trash and sewage into adjacent landfills. It is images such as these that cut through the distance. By lifting the veil that typically obscures the close connection between waste and food, they leave a suitably bad taste in our mouths (Figure 9.4).

9.5 Conclusion

Stefansson once proclaimed that "it is doubtful that man has anywhere else on earth managed an adjustment to environment so nearly perfect as that of the Eskimo to his Arctic" [9, p. 17]. Despite this recognition, he was adamant about the outside world's capacity for betterment. He coaxed the government and investors to envision the North, not as the balanced ecosystem he described above, but as a source for endless consumption: a buffet, if you will. In the ensuing rush to consume, the cacophony of cargo ships, of oil being sucked and money made, has drowned out the voices of those not invited to the table. Hunger remains at an all-time high in the Canadian Arctic. As of 2021, 76% of Inuit in Nunavut over the age of fifteen reported their household was food insecure [21]. That statistic for the rest of Canada is less than 16%. Recognizing the severity of the issue, the federal government launched the Nutrition North Canada program in 2011 to improve the affordability and availability of nutritious foods for living in Arctic communities. The program followed a market-driven approach, giving full control of the subsidies to retailers and suppliers to reduce food prices more effectively. Rates of food insecurity in the territory have increased 13% since the program's launch [22]. "I call it the Wild, Wild West because anything goes," commented one Arctic grocery employee to a national news agency in regard to the company's pricing [23].

Due in part to this price gouging, and the further scarcity it creates, there is no greater celebration in Arctic communities than that of the community feast from "country food" still gathered from knowledge and co-existence with the land. Seal, caribou, whale, and fish are spread across the gymnasium floor on cardboard surfaces, free for all to contribute to and take from. As a community event, it serves as an affirmation of a larger social body, one that ensures all its components are fed and cared for. Everyone fills themselves in unison. Yet the communal sharing of food is also a tool for self-regulation; a reminder that others are equally in need of food; and a reminder of the many eyes that are watching the food pile, ensuring that none in their midst exceed their quota, and tip the scales toward gluttony.

Akin to the mythological glutton, our era of relentless harvesting of the Arctic seems too grotesque to be represented through human form. A major challenge of the Anthropocene is precisely that it's faceless. Its magnitude obscures the role of individuals in the quickening processes of global environmental change. Like the monstrosity of folktales, complicity is buried in us all, because we rarely choose to confront it. We point to the voracious hunger of industry—fisheries, shipping, and mining feed that from the land, removing valuable sources of food and metal for redistribution overseas. We acknowledge the appetites of the extreme climate created through industry's insatiability; shorelines and glaciers literally eaten away by sun and sea. Yet the impact of these changes ultimately falls on people.

Contemporary art plays an important role in helping us see humanity in perspective to these rapid changes impacting our world. On the one hand, it helps us demonize the excessive breadth of industry, environmental ruin, and food disparity by giving them the frightful appearances that corporate and commercial interests do their best to hide. Art is a medium capable of showing how big these threats and their appetites really are, from the "Hell Mouth" depicted in Jude's sculpture *All-Consuming*, which towers above the many boats feeding it, to the monstrous fish devouring an entire world in Shuvinai Ashoona's 2018 print *Composition (Monster Eating the World)*. Yet the scale of these gluttons is always marked in relation to the humans in their midst. In Shuvinai Ashoona's print, a lone, minute hunter raises their rifle in protest. These depictions juxtapose the human and the monster, signaling that while bound through complex and often-complicit ways, the two are ultimately separate. These artworks allow us to take heart in the fact that Inuit stories do not often end well for the glutton. A reach too far, or a mouthful too many, greed is always their undoing. The ceaseless consumption of the Arctic no doubt has its own expiry date, though whether it meets its end—as it so often does in Inuit legend—through the cunning of a quick-thinking protagonist, or ultimately just eats itself out of house and home, is yet to be determined.

References

1. Stefansson V. *The Northward Course of Empire*. New York: Harcourt, Brace and Company; 1922.
2. Statistics Canada. Food security status of First Nations people living off reserve, Métis and Inuit by age group [Internet]. 2024 [cited 2025 Jan 1]. Available from: https://www150.statcan.gc.ca/t1/tbl1/en/tv.action?pid=4110006301&pickMembers%5B0%5D=1.1&pickMembers%5B1%5D=2.6&pickMembers%5B2%5D=4.1
3. Bird K, Charpentier R, Gautier D, Houseknecht D, Klett T, Pitman J, et al. Circum-Arctic Resource Appraisal: Estimates of Undiscovered Oil and Gas North of the Arctic Circle [Internet]. U.S. Geological Survey; 2008 [cited 2024 Nov 10]. Available from: https://pubs.usgs.gov/publication/fs20083049
4. Roston E. The world has discovered a $1 trillion ocean. Bloomberg [Internet]. 2016 Jan 21 [cited 2024 Dec 1]; Available from: https://www.bloomberg.com/news/articles/2016-01-1/the-world-has-discovered-a-1-trillion-ocean
5. Alvarez J, Yumashev D, Whiteman G. A framework for assessing the economic impacts of Arctic change. *Ambio*. 2020;49:407–418.

6. Hanacek K, Kroger M, Scheidel A, Rojas F, Martinez-Alier J. On thin Ice: The Arctic commodity extraction frontier and environmental conflicts. *Ecological Economics*. 2022;191.
7. Arctic Economic Council. *The State of Arctic Food: Report on the Importance of Arctic Food Systems for Local Communities and Global Needs*. Norway: Arctic Economic Council; 2023. p. 145.
8. Igloliorte H. Inuit Art is a Marker of Cultural Resilience. 2010;(Spring/Summer): 4-11.
9. Stefansson V. *The Fat of the Land*. New York: Macmillan; 1956.
10. Jenness D. The Life of the Copper Eskimos. F.A. Acland; 1922. (Report of the Canadian Arctic Expedition 1913–18: Southern Party; vol. 12).
11. Stefansson V. *The Stefánsson-Anderson Arctic Expedition of the American Museum: Preliminary Ethnological Report*. New York: Published by order of Trustees; 1914.
12. Qitsualik RA. Nunani: Glutton (Part Two). *Nunatsiaq News*. 2023 Aug 15;2.
13. Rasmussen K. *Eskimo Folk-Tales*. Copenhagen: Gyldendal; 1921. p. 156.
14. Qitsualik R. Nunani: Glutton (Part Three). Nunatsiaq News [Internet]. 2003 Aug 22 [cited 2024 Nov 10]; Available from: https://nunatsiaq.com/stories/article/nunani_glutton_part_three/
15. Qitsualik RA. Cannibal: Part Two. Nunatsiaq News [Internet]. 1999 Aug 13 [cited 2024 Nov 18]; Available from: https://nunatsiaq.com/stories/article/cannibal_part_two/
16. Qitsualik RA. Cannibal: Part One. Nunatsiaq News [Internet]. 1999 Jul 30 [cited 2024 Nov 18]; Available from: https://nunatsiaq.com/stories/article/cannibal_part_one/
17. Stefansson V. *The Friendly Arctic*. New York: The Macmillan Company; 1921.
18. Jenness D. The cultural transformation of the Copper Eskimo. *Geographical Review*. 1921;11(4):541–550.
19. Truth and Reconciliation Commission of Canada. *Honoring the Truth, Reconciling for the Future: Summary of the Final Report of the Truth and Reconciliation Commission of Canada*. Ottawa: Truth and Reconciliation Commission of Canada; 2015.
20. Mosby I, Galloway T. 'The abiding condition was hunger': Assessing the long-term biological and health effects of malnutrition and hunger in Canada's residential schools. *British Journal of Canadian Studies*. 2017;30(2):147–162.
21. Nunavut Food Security Coalition [Internet]. 2024 [cited 2024 Nov 18]. Rates of Food Security. Available from: https://www.nunavutfoodsecurity.ca/en/rates#:~:text=As%20of%202021%2C%2076%25%20of,Canada%20is%20less%20than%2016%25
22. St-Germain AAF, Galloway T, Tarasuk V. Food insecurity in Nunavut following the introduction of Nutrition North Canada. *Canadian Medical Association Journal*. 2019 May 21;191(20):E5522–E5528.
23. D'Souza S, Guerriero L, Ellenwood L, Angelovski I. What's behind rising food costs in Canada's North? CBC News: Eye on the Arctic [Internet]. 2024 Feb 23 [cited 2024 Dec 1]; Available from: https://www.rcinet.ca/eye-on-the-arctic/2024/02/23/whats-behind-rising-food-costs-in-canadas-north/

Part III

Artistic Practices That May Deepen Connection or Reconnect People or Communities with the Environment and Cultural Heritage

10 To Heal the Woods of the Earth (and the Mind) through Play and Art

Antti Stöckell and Nina Luostarinen

10.1 Introduction

10.1.1 Developing Sustainable Services in the Forest

The intangible and tangible resources provided by forests—forest ecosystem services—are a crucial foundation for tourism and wellness services in Lapland. Concurrently, tourists are increasingly aware of threats to forests, such as climate change, biodiversity loss, overlogging, as well as the wearing of terrain in tourist areas, accelerating the need to redesign services related to forest resources. The pandemic has also highlighted new needs, particularly for diversifying tourism services that can be conducted safely with small groups.

We examine the art-based activities conducted during the Forest Ecosystem Services as Sustainable Potential in Creative Industries project from the perspective of adult play, asking how can play serve as an inspirating force in a sustainable integration of forest nature, culture, and art-based activities when developing new services and activities? Play was not in the project plan and at the center of the services to be developed; instead, project plan is talking about art-based services. We share a common interest in the potential of play and examine the playful features of the project's art-based activities, in order to create a foundation for a stronger consideration of play in similar contexts in the future.

We see strong connections between art, playfulness, and being in the forest. Various definitions of art refer to creativity and freedom, unpredictable processes where there are no specific correct or rational solutions. In this sense, artistic activity already has a connection to the world of play. Interest in adult play is also growing, and we can even speak of a playful turn [1, 2]. The forest environment, central to the project, is known to liberate adults from behavior patterns guided by everyday norms: in the forest, one is allowed to relax control and free oneself from various obligations and expectations that otherwise surround daily life [3].

According to Lehtonen [4], in the midst of sustainability crises, we need art education, creative play, and a sense of drama to achieve new ways of seeing and thinking. Drawing on Eernstman and Wals Lehtonen emphasizes how artistic work at its best gives space to wonder in multiple voices and to imagine

DOI: 10.4324/9781003545880-14

and feel freely. Collaborative artistic work around environmental issues can, she argues, enable experiences of meaningfulness, comfort, and compassion. Play and communal creative activities help to imagine different realities and bring hope [4].

Our goal is to inspire people to realize the power of play as an enlivening energy in art-based services connected to the forest and as a one thread in sustainable nature connection. In times of worrying environmental changes, it is essential to create activities, content, and meanings that foster joy and hope. We hypothesize that adult play can have unique potential in this endeavor.

10.1.2 Forest Ecosystem Services as Sustainable Potential in Creative Industries Project

In the Forest Ecosystem Services as Sustainable Potential in Creative Industries project (2021–2022) conducted by the Faculty of Art and Design at the University of Lapland, art-based services responding to identified needs were conceptualized. The project was funded by the Regional Council of Lapland's appropriation for sustainable growth and vitality of regions. Stöckell served as the project manager and artist, while Luostarinen was an artist in one workshop.

In the project's service concepts, the forest, people, and art intersected in various ways. The experiments, referred to as forest encounters, were conducted in the Lapland region in collaboration with artists, tourism and wellness entrepreneurs, and work groups participating in workshops. Activities were carefully adapted to the context: alongside old forests on hiking routes, the value of brushwood in felled forests was equally recognized, finding the forest's worth even where it does not flourish at first sight. In addition to considerations of sustainable natural use and pandemic-era safety, the activities were guided by respect for Northern ways of life, referring to both Indigenous cultures and all sustainable ecoculture [5].

The project collected a diverse set of research materials typical of art-based action research [6], with the primary material being the production created during the development of seven different service concepts. The project results are presented in the Finnish publication "Art-Based Forest Encounters" [7], which serves as the starting point of our current chapter [8].

The project consisted of several experiments, of which we will mainly look at those that we designed and implemented ourselves. In the experiments conducted by Stöckell, participants spent the night in the forest in different ways at different times of the year, such as under the traditional half-open canvas shelter in autumn and in a quinzee (snow cave) in winter. The shelter provided by the forest can be seen as a cultural ecosystem service, as well as an opportunity to experience the way animals shelter themselves. The art-based activity seized the opportunities offered by fire and light and snow.

In the workshop designed by Luostarinen, participants worked with a tourism and program services entrepreneur to restore a drained swamp on land owned by the entrepreneur. The stunted pine trees that had grown on the swamp due to drainage were cut down and made into wood chips. The wood chips were used to create environmental art to celebrate the marsh's reversible transformation.

10.1.3 The Four-Field Model of Forest Play

We structure our text according to Figure 10.1 into four interrelated areas, which we have named with the letters P, L, A, and Y. The purpose of the model we created is to playfully visualize the entity we are dealing with—this is the ball we are bouncing and throwing, come along!

We start in our four-field model from play (P), moving toward the forest of natural science (L) and culture (A). In field Y, all fields combine. The facts of natural science and the freedom of play, as well as nature and culture, are not opposing, separate extremes, but a continuum blending into each other. In our model, diagonally intersecting lines are important. The natural science facts (L) and the cultural forest (A) are not opposite, separate extremes, but a merging continuum. The play and art related to the forest intersect at the junction of the cultural and scientific forest, not separating them, but restoring and renewing a sustainable link.

P = Play

Field P dives into the world of play. Amid contemporary threats, play offers a necessary therapeutic way to examine our role and participation in ecosystems. Play can provide an alibi [9], an acceptable reason, to act differently.

L = Lushwood

Field L represents the foundation, which is the forest itself, viewed natural scientifically as an ecosystem. The branches of natural science focusing on the forest provide rich descriptions of ecosystems' diversity and their

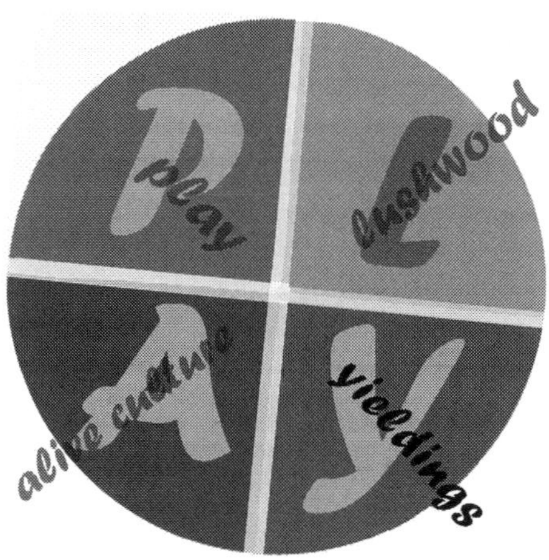

Figure 10.1 The four-field model of art-based forest play. Illustration: Antti Stöckell 2024.

fragile vulnerability. Scientific knowledge is one significant root for both art and play in the forest.

A = Alive Culture

In field A, we arrive in the cultural forest. The project explored new ways to earn a livelihood from the forest, aiming to identify especially the still latent growth areas of the forest's cultural ecosystem services. These are based on the same ecosystems that maintain the forest's biodiversity, emphasizing the sustainability of art-based activities.

Y = Yieldings

In the fourth field, we summarize by integrating all the previous fields, summarizing the answer to our question of how play can inspire in the context of combining forest nature, culture, and art-based activities in the sustainable development of new services.

10.2 Connections between Play, Art, and the Forest (Field P, Play)

Several researchers of play [10, 11] say that play creates special place and time that deviates from normal reality, when the normal rules and routines of communities also cease to exist. Meanings and roles can be played with, located in between of different realities, in a liminal space that frees people from structural obligations [12, 13]. The creative chaos of the liminal space makes familiar things less familiar, enabling reinterpretation and new perspectives [14].

However, play has its boundaries and rules. In the forest, it must be sustainable for the ecosystem and safe for the players. The thousands of small species mentioned earlier are both real and require imagination due to their invisibility. Thus, forest play does not need to reach beyond reality but can make the invisible visible and understandable through play. For example, the snow hole of a wintering bird is not invisible, so it is too small for humans to experience directly. By building a suitable-sized shelter and spending the night in it like a bird, one can genuinely feel the protection offered by snow on a frosty night (Figure 10.2). Playing in the snow often evokes childhood memories and liberates adults into playfulness. Building the shelter is also a serious survival skill practice, during which the threshold for adult play can suitably lower.

Calling an art project or workshop activity "play" aims to lower barriers to participation; creating art might seem demanding, but calling it play can encourage participation. However, using the concept of play might need justification, as adult play is often seen as stigmatizing and taboo. Play researcher Katariina Heljakka states in an interview with Väki magazine that play can be seen as regression, and in a society that values efficiency, it can detract from credibility and seriousness [15]. Laura Seesmeri [16] and her colleagues note that while play benefits all ages, adults seek reasons, benefits, and purposes for their actions—things not inherently linked to play. They argue that art, which has a purpose in the adult world, can momentarily lead adults into play.

In an impressive natural environment, we are also more tolerant of others' unusual behavior, different from the normal adult world [17], making the forest

Figure 10.2 Captions clockwise from top left to bottom left: (Top left) Examining the age of the forest is a current topic and is strongly related to the discussion about the use and protection of forests in Finland. How old is the spruce, under whose branches three hikers can sleep, and whose branches are like sturdy ladders for a playful climber? Photograph by Antti Stöckell, 2022. (Top right) When a healthy young person crafts a walking stick, there is a touch of play, which becomes game-like when ten different tree species are sought to attach to the stick. Naming and categorizing species are typical of natural science, which can be learned in a fun and active way. The tree samples were cut from suitable thickets without damaging the vitality of nature. Photograph by Timo Jokela, 2022. (Bottom right) We have shared concerns about our living environments with forest grouses and wading birds. Will there be enough soft snow to protect ourselves in the freezing cold, and will restored peatlands become good habitats for species that need them? Playful activities that intertwine knowledge and art brought joy and hope even while contemplating concerns. Photograph by Antti Kurola 2021. (Bottom left) A quinzee was made. Photograph by Antti Stöckell, 2022.

an ideal playground for adults. Breaking norms might evoke feelings of embarrassment, but also a fun, exiting feeling of breaking boundaries. Playing in the forest allows adults to safely break norms associated with adulthood, potentially providing great pleasure.

We base our idea of the connection between art and play like Venäläinen [18] on Dissanayake's thought from the 1970s about the similarities between art, play, and ritual. Dissanayake highlights features typical of art: freedom, unpredictability, imagination, invention, and joy.

Perspectives on the connection between forest and play can be found in Gadamer's ideas presented by Jylhämo [19]; play is nature's play in man; play repeats the idea of nature's self-sufficiency, happening by itself, like nature. Play does not require that there should be a subject who plays, because play only achieves its goal when the player disappears into play [19]. A significant part of the magic of play is related to this kind of liberating dispersal of one's own subjectivity. However, it is not possible to impose a condition on players that they must lose their subjectivity, as voluntariness is always a basic value of play. We can only try to create conditions for the free realization of play.

Gadamer also connects artworks to the autonomy of play and nature, asserting that "human play achieves its fulfillment in art" [19, p. 83]. Frans Mäyrä [20], in his historical overview of play and game studies, notes that German writer and philosopher Friedrich Schiller (1759–1805) emphasized the significance of play and game as an integral part of human activity and culture, where we most authentically realize ourselves.

Piironen [21] examines concept art, land art, process art, and performance art from the perspective of play, highlighting sand, soil, and mud as materials that are central also in children's play. The use of the body as a tool for play and artistic creation is also natural: the body is wrapped, covered, and dyed with earthy materials, and one digs into sand or soil. Piironen suggests that using found materials as tools for play can be seen as a cognitive process of exploring the environment and the properties of materials. Finding the interpretive idea of an object or material as the starting point for play is analogous to conceptual art, which primarily seeks to make thinking visible. Piironen [22] also notes how the central content in land art, for example, seems to be a return to childhood play and the unconscious. According to her, the use of these natural materials and actions points to primitive thinking and our collective subconscious. The intertwining of the materiality and corporeality of elements in the forest ecosystem can be found or imagined in ideas like: a burning stick draws a fiery arc as a person waves it through the air, or a person throws wood chips into the air and becomes enveloped by them. Moments that are fleeting yet meaningful, as cognitive, conceptual, and physical activities merge into acting and knowing with material, circumstances, and place.

Mäyrä [20] notes that the concept of ludification, or the playful turn, describes a broader shift toward a more prominent role of play and playfulness in various aspects of contemporary society. He suggests that ludification can be seen as a positive phenomenon that increases people's freedom to act and express themselves. Frissen et al. [1] highlight "playful identities," the trend of flexibility and playfulness in people's self-image and self-presentations, which has a broad impact on Western media and culture. From this perspective, we see that play can offer a different form of freedom for creative action and self-expression than art, liberating individuals to act without the pressure of results.

Adult play invitations aim to facilitate, for example, the processing of difficult issues, instead of blame, by bringing joy and reward as the underlying tone of methods. As Rebecca Solnit [23] reminds us, it is important to maintain hope. If one is too optimistic or falls into despair, nothing gets done. According to Panu Pihkala [24], various methods that utilize bodily movement, encounters, creative expression, and physical connection to other beings play a significant role in processing environmental emotions and fostering hope. Henna Laininen [25] echoes Pihkala's thoughts stating how the dimension of joy and play inherent in art can provide experiences of endurance during times of crisis. The power of artistic expression lies in its multiplicity, playfulness, and unpredictability, far from the declaration of simplistic truths. Whether one sees dimensions of play in art or artistic means in play is a matter of perspective.

Luostarinen [26] notes how playing requires an escape from reality into a space where different laws apply; the unusual becomes ordinary, and objects and people can be given unusual attributes and names. Through play, players take control of the space and content of play, adopting roles and behaviors that often deviate from the norms of everyday reality. Her artistic interventions, conducted as part of his research, showed that the alibi of play, by creating a bubble of permissible play, can catalyze the full realization of adult play. Simultaneously, play expands into a part of place experience, and a light layer of play is drawn over everyday life.

In the interventions of Luostarinen's research, the goal was the realization of play [26]. Although the goal of the experiments in our forest project was not play, our data shows that in the forest, the transition to playful moments that differ from the norms of the real world happened almost unnoticed. At the end of the very physical work of Liro's figure, it was natural to lie down on the soft ground like a bird's feathers. Inside the snow cave (Figure 10.2, number 3), on the other hand, just the thought that this way the bird can protect itself against the cold in winter was enough. The previous joint work freed the participants from social norms of behavior, and the created figure of a bird, or a snow cave, provided an alibi for playfully empathizing with the status of another species.

Hannula [27] characterizes the philosophy of play, describing how its aim is to encounter and open up, to let things happen without coercion. In the Forest Ecosystem Services project, no ready-made services were created; instead, possibilities were explored. Thus, Hannula's continuation of the philosophy of play fits well: instead of precise productization, "what was and what is to come, what we want, fear, desire, and hope for…" are brought together into the sphere of interaction [27, p. 13]. This also applies to the philosophy of forest play, as the perspective emphasizing the past and the future seems to naturally relate to the temporality of the forest, as well as our fears and hopes related to issues like biodiversity loss. The desire, listed by Hannula, on the other hand, manifested itself in the project as a direct action toward the desired future.

10.3 The Hidden Abundance of Northern Forest Ecosystems (Field L, Lushwood)

The forests in the project's implementation area near the Northern polar circle do not appear lush nor teeming with diverse life at first glance; much of their biodiversity is hidden. Quantitative counting and classification, typical of natural science, are necessary to understand the forest's abundance. It is estimated that over 20,000 different species live in Finnish forests, the majority of which are insects or other invertebrates, as well as various fungi and algae. About 2,500 species use living plant parts for food, 4,000 species live as parasites on other species, and 4,000–5,000 species live on decaying wood [28]. Thus, a key aspect of forest biodiversity manifests in small and often hidden forms. In the soil, where half of the forest's growth occurs, a network formed by roots and fungi saturated with bacteria intertwines—in one gram of soil around the roots, there can be 10 billion bacteria [29]. Besides the small size of the most abundant species, another notable feature is time—insects and fungi flourish only briefly in easily noticeable forms, spending the rest of the time as eggs, larvae, and spores, almost invisible. The creation of deadwood, invaluable for biodiversity, is very slow. Forest life thus follows continuous cycles of change, varying greatly in duration.

In the forest, the functional entity formed by the interacting organisms and their inanimate environment is called an ecosystem [30]. An ecosystem can be identified by its structure, composition, and processes, and further hierarchically, the organism community, population, and individuals [30]. Ecosystem services are the benefits produced by ecosystem functions for the rest of the ecosystem or for humans [30]. Each of the tens of thousands of species has its place and function in ecosystems and their interconnected networks.

From the human perspective, trees make the forest a forest. The difference between commercial forests and protected old forests can be dramatic, particularly regarding the age structure and number of species. The oldest living pine in Finland began its growth in the 13th century, and pines typically live to be 300–400 years old [31]. In commercial forests, which make up the majority of our forests, trees are usually harvested at 80 years old, which is roughly the average lifespan of Finns. Ensuring forest biodiversity is widely discussed and creates pressure to critically examine forestry methods. Art-based activities can play an important role in bringing constructive perspectives to the discussion.

Does focusing on problems undermine the development of experiential services? Can forests be viewed as species and processes without considering ecological crisis issues? Simply examining species and emphasizing their protection can lead well-intentioned activities astray. The concept of ecosystem services aims to dissolve the opposition and conflict between nature conservation and its use [30]. The Wood Sandpiper workshop (Figure 10.2), conceptualized by Luostarinen, was based on research data about the consequences of peatland drainage and restoration. When a local tourism service provider aimed to restore a previously drained peatland area by blocking ditches and removing trees, the project saw an opportunity for art-based activities. A group of office workers was invited to spend a well-being day on the peatland, where they shaped the felled and chipped wood into a large Wood Sandpiper shape, one of the species benefiting from the restoration. The Wood Sandpiper would invite its kin to return. The workshop turned concern into action, generating joy and hope. As the peatland is restored, its crucial water cycle regulating ecosystem service can gradually recover. At the same time, the restored area invites species favoring open spaces and enhances the aesthetic landscape experience for visitors. Thus, the tourism entrepreneur benefits from the improved cultural ecosystem services of the restored peatland. Figure 10.2 of the Wood Sandpiper will soon decompose back into the peatland, as it should be in an ecosystem. The presentation of the workshop and its results were enthusiastically received by a variety of audiences, including forest industry stakeholders, when the Liro workshop was presented at the Lapland Forestry Days to 400 industry professionals.

10.4 Culture Lives from the Forest (Field A, Alive Culture)

According to Hankonen [32], who has studied the experiences of people who spend a lot of time in nature for leisure and various work, nature is not a clearly defined category but rather a "diverse compilation of cultural meanings-bearing places and elements and culture-producing agents" [32, p. 248]. Thus, culture cannot be separated as distinct from nature. Hankonen describes the blending of systems of meaning and the blending of those in nature with various networks of actors as an existential forest, to which she amusingly applies the natural science concept of mixed forest. In this culturally mixed forest, meanings from forestry, nature conservation, aesthetics, folk and hiking cultures, as well as perceptions derived from art and religion, blend together to form ideas about the character of the place. "In the mixed forest, the human and the non-human blend together; humans are part of nature and produce meanings together with nature" [32, p. 248]. Thus, the forest is not purely natural, nor are traces of human activity found there purely cultural.

In gathering, fishing, and hunting cultures, livelihoods have relied on the yield of the forest, i.e., ecosystem services provided by the forest ecosystem. Stories, myths, and beliefs reflect past generations' often respectful relationship with the provider of livelihood and, on the other hand, the consequences of lacking that respect. Elements identified as culturally historical in the forest, such as carvings on the trunks of old trees and various decomposing wooden structures and remnants of buildings, presented by Taivainen [33], are intriguing for art-based services. They offer a fascinating journey through time and an opportunity to examine humans as part of nature, interacting with various elements of the forest. These traces do not need to be sought out in protected areas; they can be found everywhere.

Traces of past generations' activities can be examined, following the approach of cultural anthropologist Ingold [34], through the concept of taskscape, as traces and remnants shaped by dwelling-related activities. Ingold's concept of taskscape also integrates nature and culture. Humans do not shape nature; rather, taskscape emerges and evolves as humans resonate with natural cycles [34]. This is precisely how we act when we follow the annual cycle, for example, by picking berries and mushrooms in the forest. Cultural ecosystem services, Hankonen's mixed forest, and Ingold's taskscape thus share the same roots and help to dismantle the opposition of conservation/use, nature/culture pairs. These concepts do not reduce the need to assess the sustainability of various activities in the forest. When utilizing ecosystem services, a key criterion for sustainability is to examine the overall balance among them—exploiting one resource should not significantly undermine others.

The merging of the concepts of nature and culture is also sought through the concept of Northern knowledge, which, according to Huhmarniemi and Jokela [5], involves local ecocultures and the knowledge formed in relation to traditions among both Indigenous and non-Indigenous peoples. They state that ecoculture is built on a combination of local cultural heritage, tacit knowledge, manual skills, and material understanding intertwined with it. In the concept of ecoculture, the unity of natural ecosystems and cultural activities is also conceptually inseparable. In ecoculture, handicrafts and art are not separated but are intertwined with each other and with daily life, being an integral part of it. This also provides a bridge toward play, which is also attached to the moment and to the offerings of nature and culture that are at hand. In this way, play can provide an important link strengthening and enlivening the connection and unity between nature and culture.

Examining the cultural history of the forest leads to the trails of timeless myths, ancient gatherer-hunter stories, and the toil of heavy forestry workers of the last century. As we enter the modern age, machines increasingly perform labor, such as the wood chipper in the Wood Sandpiper workshop, efficiently, and humans have the opportunity for play, for example, in the form of increased leisure time. Thus, taskscape can be transformed into playscape or play-escape [35], where escaping to the playful forest interrupts the perhaps monotonous daily grind. Play can also incorporate stories of survival struggles and toil. A few hours of clearing work in the workshop evoked memories of youth in older participants and allowed them to empathize with the hard work of previous generations. The brief toil was like play compared to weeks and years of labor. This playful moment activated the tacit knowledge stored in the body, which was also naturally verbalized in the situation. Now, physical labor had a new meaning and goal compared to previous experiences. The transformation of the peatland was experienced by settling, as into feathers of the big wading bird (Figure 10.2), inspired by the transformation stories of the Kalevala. In such situations, even visitors can learn from the Northern knowledge and ecoculture described above. Learning-enhancing services are essential in the tourism and wellness industry.

In art education, the forest has been seen as a significant learning environment, and forest play as one way to address sustainability issues [36, 37]. Children have played in nearby forests throughout history. Children's ability to create a world of play from the forest and its elements seems almost innate if they are given the opportunity to play in the forest. In other words, children also utilize the forest's cultural ecosystem service in their play. In hut and nest building plays, human and animal dwellings merge. In two forest encounters of the project, shelters and nests were built creatively using terrain features and seasons.

It's fascinating to think about how a nature element that fuels the imagination inspired someone's play hundreds of years ago and the same element still inspires play today. In adult forest play, there may be similar elements; instead of a small nest, the whole forest can be a lap to which even adults can climb and be small.

10.5 Play Weaves Together Knowledge and Feeling, Us and the Forest (Field Y, Yieldings)

We have progressed in our four-field model (Figure 10.1) by examining the possibilities of play in the forest, emphasizing in turn the perspectives of natural science and culture. Now, let's crystallize the answer to our question: how can play serve as an inspiring force in a sustainable integration of forest nature, culture, and art-based activities when developing new services and activities? From the project's perspective, we think specifically of tourism and wellness entrepreneurs collaborating with artists to develop their services, as well as the users of these services, the customers. Expanding the possibilities of art-based activities that understand the dimensions of play in forests in a sustainable way naturally also serves all non-commercial activities, especially in art and environmental education.

How does play inspire? Initially, one must overcome the threshold of play, which may have grown high for adults. Once over it, anything is possible. When one norm, "adults don't play," is broken, other norms can be broken as well. Since play resides in the borderlands of different realities, the normal rules are not allowed to restrict the search for new perspectives. Play cannot be forced. Flexible plans and loose schedules can create an immediate atmosphere where play can occur spontaneously and in small moments. In these moments, the significance of action and new possibilities can unfold in surprising ways.

What factors contribute to sustainability? In a holistic approach that combines nature and culture, even difficult questions are not ignored but considered part of play, including concerns and problems caused by human activity. This can succeed when understanding of the forest ecosystem and humans as part of it is fostered through the joy and functionality of play.

How do the forest, culture, and art combine? In our four-field model (Figure 10.1), play and art intersect at the confluence of the cultural and natural forest, not separating them from each other, but restoring and renewing a sustainable connection. Those developing new activities in the forest need to recognize their place and relationship in the mixed forest formed by various human and non-human actors. The foundation of intangible and tangible resources in the forest is the same ecosystem. The vitality of the forest living culture depends on the vitality of the forest ecosystem.

Art-based forest play takes place on a mixed, layered ground of nature and culture, which in the era of ecological crises also bears scars from human activity. Play is not just a bandage for wounds; it has the potential to promote the well-being of both humans and nature. Seen through the eyes of art-based play or play-based art, a previously disregarded wasteland may appear useful and even renewable resource of a cultural ecosystem service that can be strengthened in the art-based activity being developed.

Art-based activities with playful moments can be small and local, strengthening the participants' connection with nature. In the face of global environmental change, it is crucial to share the good news through a wide range of publications and exhibitions, to tell of even small successes and the joy experienced in the forest in a way that inspires hope and encourages action. We believe in the enchantment and power of play. As we are encouraged to play in the embrace of the forest and in the name of art—or to make art in the name of play—we can learn about ourselves and each other, as well as about our connection with the forest.

References

1. Frissen V, De Mul J, Raessens J. Homo Ludens 2.0: Play, media and identity. In: Thissen J, Zwijnenberg R, Zijlmans K, editors. *Contemporary Culture: New Directions in Art and Humanities Research*. Amsterdam University Press; 2013. pp. 75–92.
2. Sutton-Smith B. *The Ambiguity of Play*. Harvard University Press; 1997.
3. Hallikainen V, Sievänen T, Tuulentie S, Tyrväinen L. Luonto kokemusten ja elämysten lähteenä. In: Tyrväinen L, Kurttila M, Sievänen T, Tuulentie S, editors. *Hyvinvointia metsästä*. SKS; 2014. pp. 36–47.
4. Lehtonen A. Kestävyyskriisien keskellä tarvitaan taidekasvatusta, luovaa leikkiä ja draaman tajua [Internet]. 2022; 53(2):203–207 [cited 2024 Nov 8]. Available from: https://journal.fi/kasvatus/article/view/115928
5. Huhmarniemi M, Jokela T. Arctic art and material culture of Arctic: Northern knowledge and cultural resilience. In: Heininen L, Exner-Pirot H, Barnes J, editors. *Arctic Yearbook 2020: Climate Change and the Arctic: Global Origins, Regional Responsibilities?* Arctic Portal; 2020.
6. Jokela T, Huhmarniemi M. Taideperustainen toimintatutkimus soveltavan taiteen kehittämisen välineenä. In: Jokela T, Huhmarniemi M, Paasovaara J, editors. *Luontokuvaus soveltavana taiteena*. Lapin yliopisto; 2020. pp. 39–61.
7. Jokela T, Stöckell A. Taideperustaisia metsäkohtaamisia: ekosysteemipalveluista elinvoimaa luoville aloille. Lapin yliopiston taiteiden tiedekunnan julkaisuja. Sarja C. Katsauksia ja puheenvuoroja. 2022;(65). Available from: http://urn.f/URN:ISBN: 978-952-337-317-4
8. Stöckell A, Luostarinen N. Lupa leikkiä metsän sylissä: taideperustaisten metsäkohtaamisten tarkastelua aikuisen leikin näkökulmasta. In: Jokela T, Stöckell A, editors. *Taideperustaisia metsäkohtaamisia: ekosysteemipalveluista elinvoimaa luoville aloille*. Lapin yliopiston taiteiden tiedekunnan julkaisuja. Sarja C. Katsauksia ja puheenvuoroja. 2022;(65):23–31. Available from: https://urn.fi/URN:ISBN: 978-952-337-317-4
9. Deterding S. Alibis for adult play: A Goffmanian account of escaping embarrassment in adult play. *Games and Culture*. 2017. Available from: doi:10.1177/1555412017721086
10. Huizinga J. *Leikkivä ihminen: Yritys kulttuurin leikkiaineksen määrittelemiseksi*. Werner Söderström; 1984.
11. Stenros J. In defence of a magic circle: The social and mental boundaries of play. *ToDiGRA*. 2014;1(2). Available from: doi:10.26503/todigra.v1i2.10
12. Kontinen T, Houni P, Karsten H, Toiviainen H. Liminaalitilan käsite työn muutosten jäsentäjänä. *Aikuiskasvatus*. 2013;2013(4):252–261. Available from: doi: 10.33336/aik.94057

13. Turner V. *From Ritual to Theatre: The Human Seriousness of Play*. PAJ Publications; 1982.
14. Mainemelis C, Ronson S. Ideas are born in fields of play: Towards a theory of play and creativity in organizational settings. *Research in Organizational Behavior*. 2006;27:81–131. Available from: doi:10.1016/S0191-3085(06)27003-5
15. Karakoski M. Leikkiä ikä kaikki. Väki. 2021;4/2021:10-14. [Internet]. 2018. [cited 2024 May 2]. Available from: https://issuu.com/vaki-mana/docs/v_ki4_issuu
16. Seesmeri L, Mellanen L, Pässilä A. Taide+tiede=! Manifesti taidelähtöisistä menetelmistä. In: Pässilä A, Parjanen S, Aarrevaara E, Laakso H, Knutas A, Harmaakorpi V, editors. *Yhteinen lähiö – kokeilut ja asukastieto kaupunkien kehittämisessä*. LUT-yliopisto, LUT Scientific and Expertise Publications/Tutkimusraportit - Research Reports 145; 2022. Available from: https://urn.fi/URN:ISBN:978-952-335-878-2
17. Sawada K, Nomura M. Influence of positive and threatened awe on the attitude toward norm violations. *Frontiers in Psychology*. 2020;11:148. Available from: doi:10.3389/fpsyg.2020.00148
18. Venäläinen P. *Nykytaide oppimisen ympäristönä: Näkemyksiä nykytaiteesta, oppimisesta ja niiden kohtaamisesta*. Jyväskylän yliopisto; 2019. Available from: http://urn.fi/URN:ISBN:978-951-39-7848-8
19. Jylhämö K. Miksi hermeneutiikan leikki pitää ottaa vakavasti? *Niin & näin: filosofinen aikakauslehti*. 2002;9(3):82–90. Available from: https://netn.fi/sites/www.netn.fi/files/netn023-23.pdf
20. Mäyrä F. Pelikulttuurien tutkimuksen historiaa. In: Friman U, Arjoranta J, Kinnunen J, Heljakka K, Stenros J, editors. *Pelit Kulttuurina*. Vastapaino; 2022. pp. 35–62.
21. Piironen L. *Leikkivä taiteilija – leikkivä lapsi: pohdintaa taiteen ja leikin yhteyksistä taidekasvatuksen kontekstissa*. [Licentiate thesis]. Taideteollinen korkeakoulu; 1998.
22. Piironen L. Leikissä taiteen ainekset. In: Piironen L, editor. *Leikin pikkujättiläinen*. WSOY; 2004. pp. 80–89.
23. Solnit R. *Hope in the Dark: Untold Histories*. Nation Books; 2006.
24. Pihkala P. Eco-anxiety and environmental education. *Sustainability*. 2020; 12(23):10149. Available from: doi:10.3390/su122310149
25. Laininen H. *Taiteen metsittymisestä – Harjoitteita jälkifossiilisiin oloihin*. Taideyliopiston Kuvataideakatemia; 2018.
26. Luostarinen N. *Paikkaleikkejä ja leikin paikkoja: Leikillisillä taideinterventioilla kohti paikkaempatiaa*. Lapin yliopisto; 2023. Available from: https://urn.f/URN:ISBN:978-952-337-392-1
27. Hannula M. *Leikin filosofia*. Parus Verus; 2021.
28. Saaristo L, Mannerkoski I, Kaipiainen-Väre H. Metsätalous ja uhanalaiset lajit. Metsätalouden kehittämiskeskus Tapio [Internet]. 2010. [cited 2024 Apr 10]. Available from: https://tapio.fi/wp-content/uploads/2019/10/Metsatalous_ja_uhanalaiset_lajit.pdf
29. Ruohonen A. Metsä maan alla. Sienet ja bakteerit pyörittävät metsän ekosysteemiä. In: Jokiranta A, Juntti P, Ruohonen A, Räihä J, editors. *Metsä meidän jälkeemme*. Like; 2019. pp. 235–237.
30. Kolström T. Mitä ekosysteemipalvelut ovat? In: Hiedanpää J, Suvantola L, Naskali A, editors. *Hyödyllinen luonto. Ekosysteemipalvelut hyvinvointimme perustana*. Vastapaino; 2010. pp. 33–51.
31. Räinä J. Turkua vanhempi mänty. In: Jokiranta A, Juntti P, Ruohonen A, Räihä J, editors. *Metsä meidän jälkeemme*. Like Kustannus Oy; 2019. pp. 32–33.
32. Hankonen I. *Ihmisiä metsässä. Luonto kulttuuriympäristökysymyksenä*. Suomen Kansantietouden Tutkijain Seura; 2021.
33. Taivainen J. *Metsien kätköissä*. Metsäkustannus; 2017.

34. Ingold T. The temporality of the landscape. *World Archaeology*. 1993;25(2): 152–174.
35. Luostarinen N, Hautio M. Play(e)scapes: Stimulation of adult play through art-based action. *The International Journal of New Media, Technology and the Arts*. 2019;14(3):25–52. Available from: https://doi.org/10.18848/2326-9987/CGP/v14i03/25-52
36. Huhmarniemi M. Metsäleikkejä kestävään kehitykseen. In: Granö P, Hiltunen M, Jokela T, editors. *Suhteessa maailmaan: ympäristöt oppimisen avaajina*. Lapin yliopistokustannus; 2018. pp. 108–132.
37. Huhmarniemi M, Hiltunen M, Jokela T. Pöheikön kutsu: taideperustaista ympäristökasvatusta pohjoisen luontokulttuurissa. In: Kyrönlampi T, Mäkitalo K, Uitto M, editors. *Esi- ja alkuopetuksen käsikirja*. PS-kustannus; 2020. pp. 199–222.

11 Blueprints and the Topography of Loss

Hannah Perrine Mode

11.1 Introduction

11.1.1 Initial Encounters with Glaciers

The ice was blue, and I was lost. I was in Iceland in 2014 for an artist residency and had decided to spend a few days solo hiking a popular route through the highlands to the coast. What began as a crowded trail had become a desolate path with piles of black volcanic ash and freezing rain, the next trail marker eluding me in the dense fog. The only detectable color outside of my cherry red raincoat was suddenly a subtle but unmistakable milky blue beneath my soaked hiking boots. I very literally stopped in my tracks, my lungs involuntarily pushed out a half-strangled *whoop* of awe.

The first time I saw a glacier was a few years prior in New York City. Although I had taken the required science classes in school, the scale of deep time and geologic forces felt overwhelming and abstract, and I felt disconnected and discouraged by these classroom learning environments. Yet every week I returned to the same bookstore in Soho, flipping through *A Portrait of Ice* by Caleb Marcus Cain. I was taken with his photographs of these strange, undulating blue forms. I eventually saved up to buy the book, and it sat on my shelf as a reminder of the vast, remote scale of landscapes I thought I would likely never encounter in person.

Then somehow, after a series of choices that had made it possible to spend two months making art in Iceland, I found myself standing on blue ice in that shallow ravine. I eventually located the next trail marker, but that *whoop* of gratitude latched itself to my body, curling up in my chest and patiently waiting for more encounters with blue ice.

11.1.2 Following Glacial Streams

With each new glacial experience in Iceland, and later the dwindling ice in Glacier National Park in Montana, my art practice shifted. Initially, I had been drawn to Iceland and subsequently the California Bay Area (where I moved post-New York), curious to explore through art-making how humans create a sense of place in specific types of natural landscapes. In my case, I was interested

DOI: 10.4324/9781003545880-15

in the link between tectonically dynamic locations – a feeling that was originally sparked when living in the Andes in Peru post-college – and the notion of "thin places": a Celtic concept that there are certain locations where the barrier between heaven and Earth is especially porous [1]. But I found myself increasingly curious about how the geology around us, not just geography, influences our emotional experiences of place, community, and ourselves.

While pursuing an MFA in Studio Art, I deepened my interest in glacial systems and human connections through material explorations of ice as an endangered commodity, documentation of climate change, and events that fostered community storytelling. With minimal prior knowledge and experience in scientific fieldwork and Earth systems science, I connected with the Juneau Icefield Research Program (JIRP) with hopes to apprentice glaciologists.

11.1.3 Fields of Ice

JIRP is an eight-week expeditionary research and educational program. Participants – mostly undergraduate and graduate students – traverse from Juneau, Alaska, to Atlin, British Columbia, staying in a series of permanent camps along the way (Figure 11.1). The program was established in 1946, and

Figure 11.1 Camp 18, one of JIRP's permanent base camps overlooking the Gilkey Glacier. Photo by Ben Huff.

JIRP has maintained the longest-running study of any glacier in the Western Hemisphere. Such comprehensive and continuous studies are rare yet crucial for understanding past and present climate change. The program now integrates art, communication, storytelling, and outreach into the broader glaciology and Earth science curriculum.

I joined JIRP in 2017 as an artist-in-residence, taking part in the full eight-week program across the Icefield with students, staff, and faculty. This experience provided me with a robust foundation in glaciology, glacier traverse safety, and field logistics and provided a platform for expanding my artwork to more directly engage with these previously foreign (now beloved) topics.

11.2 Art-Science Intersections

11.2.1 Embodied Understandings

My initial experiences with glaciers opened critical doors into other fields. Personal, embodied experience in glacial landscapes offered me the ability to learn about Earth science through my own creative lens and through place-based education and research models [2]. I also found a connection with an incredible network of people I met through this work over the years, including many long-term collaborators and friends. This baseline of knowledge and connection then offered an opportunity for more work in similar fields to develop and implement interdisciplinary art-science integration.

I have spent many summers working as an artist and educator in glacial environments in Alaska, an area where alpine glaciers are melting at some of the fastest rates on Earth [3]. In addition to JIRP, I have also worked with Inspiring Girls* Alaska (Girls* on Ice), Inspiring Girls* Aotearoa New Zealand (Girls* on Ice), Chulengo Expeditions in Chile and Alaska, and the International Arctic Science Committee in Iceland. Observing the impacts of the changing climate in these places has made me particularly passionate about applying interdisciplinary approaches to polar science.

11.2.2 Creative Connections

Today I work at the confluence of art, environmental science, and communication to explore how these interrelated fields are complementary to foster open exchange across disciplines. I view art and science as entangled domains rooted in curiosity, experimentation, and communication – both subjects benefiting from a multiplicity of approaches, methodologies, and frameworks. My research and practice examine how creative research and communication can enhance engagement with polar science, promoting inclusive, place-based, and diverse perspectives through alternative forms of knowledge production [4, 5].

Across all my work, I harness creative practices to reveal and deepen human connection to Earth systems, expanding methods of art-science research, co-creation, and communication. I am passionate about revealing links between humans and geological forces through creative methods, fostering empathy, storytelling, and environmental stewardship. Approaching glaciated landscapes through a creative lens allows for a deeper, more nuanced understanding of the environmental systems that I am a part of. Experiencing the love, loss, and transformation inherent in learning about these landscapes is a profound privilege.

My goal is to advance science communication through my research, developing methodologies for sustained collaboration between creative and scientific disciplines. As part of this work, I model interdisciplinary methodologies by interweaving my art practice with educational programs and research. This interdisciplinary approach strengthens critical systems thinking [6] while promoting a more inclusive understanding of Earth science by valuing alternative forms of knowledge production and learning styles [7].

I return annually to JIRP as Teaching Faculty and volunteer year-round on the Academic Council leadership team. I develop science communication curricula and creative frameworks for understanding climate and glacier science. I have experienced how art, especially when generated during the research process and in collaboration with scientists, increases entry points for those with different learning styles and backgrounds to participate in science. These diverse entry points help lower barriers to scientific participation. There are many scientists, artists, researchers, and educators at the forefront of art-science integration, and I consider myself fortunate to work alongside and to be continually inspired by those dedicated to this work.

11.3 Climate Communication in Context

As we confront a global crisis, scientists emphasize the urgent need for action to mitigate the most extreme impacts of human-caused climate change [8], creating spaces and frameworks for interdisciplinary and collaborative work is crucial. In our current climate crisis, creative practices can be critical to amplify knowledge production, diversify learning, and increase communication in polar science education and research, both internally and externally. It is essential to understand and employ innovative methods that deeply and creatively engage scientists and members of society alike.

There are significant challenges in communicating climate science and glacier retreat, as well as barriers to public understanding of these complex topics [9]. Yet art has the potential to deepen personal connections and ease political division around climate change [10, 11] by providing unique opportunities for learning, reflection, empathy, and discourse. Throughout my work with climate and glacier science, I have developed a visual framework for better understanding the ways in which art may be integrated into the scientific process at all stages, not just as a tool for outreach; while science illustration is an important component, it is not the only valid method of science communication and art-science collaboration.

In the next section, I will discuss how integrating creative practices into a polar science context, specifically my work with JIRP as an artist and educator, can provide unique pathways for exploring and understanding environmental science, enriching the study of this unique and rapidly changing environment. This interdisciplinary approach fosters a more comprehensive understanding of glacial environments and the complexities of climate systems.

11.4 Integration Framework

There are many opportunities throughout the scientific research process that welcome – and benefit from – the integration of creative practices. I have identified four steps in this process and designed a visual that highlights the importance of art-science integration along the science learning and research journey: Data Collection, Synthesis, Outreach, and Participation.

Internally, in the Data Collection and Synthesis stages, creative learning practices increase research opportunities and methods [12]. Externally, in the Outreach and Participation phases, creative communication increases access to scientific ideas [13] and, in turn, makes science concepts more accessible to new audiences.

11.4.1 Data Collection

Collaborative research is a significant area where art contributes to Arctic science. Through interdisciplinary projects that blend artistic practices with scientific inquiry, researchers can explore complex environmental data and phenomena in novel ways [14]. This collaboration leads to richer, more nuanced understandings of Arctic systems, as art provides new tools for visualizing and interpreting scientific findings.

11.4.2 Synthesis

Artistic techniques also offer unique frameworks for synthesizing complex data and abstract information [15]. By visualizing and interpreting Arctic landscapes through creative media, scientists can present their findings in more accessible and engaging formats. This not only aids in the communication of complex scientific data but also enhances public understanding of Arctic environmental issues.

11.4.3 Outreach

Effective communication of Arctic science is further enriched by artistic expression [13]. Visual art, storytelling, and multimedia presentations capture the nuances of Arctic environments and alternative depictions of climate change in compelling ways. These creative forms make scientific research more relatable and impactful, helping to engage a broader audience in the critical issues facing the Arctic.

11.4.4 Participation

Art enhances inclusive access to Science, Technology, Engineering, Art, and Mathematics (STEAM) subjects by accommodating diverse learning styles and backgrounds [16]. Artistic methods make polar science more approachable, enabling students and researchers to engage with the subject matter in a way that resonates with their personal experiences and perspectives. Creative practices thus bridge educational gaps and promote broader participation in scientific inquiry (Figure 11.2).

11.5 Case Studies in Creative Fieldwork

I have chosen to highlight six of my projects – three art-making and three teaching – related to my work with JIRP in Alaska and mapped where they

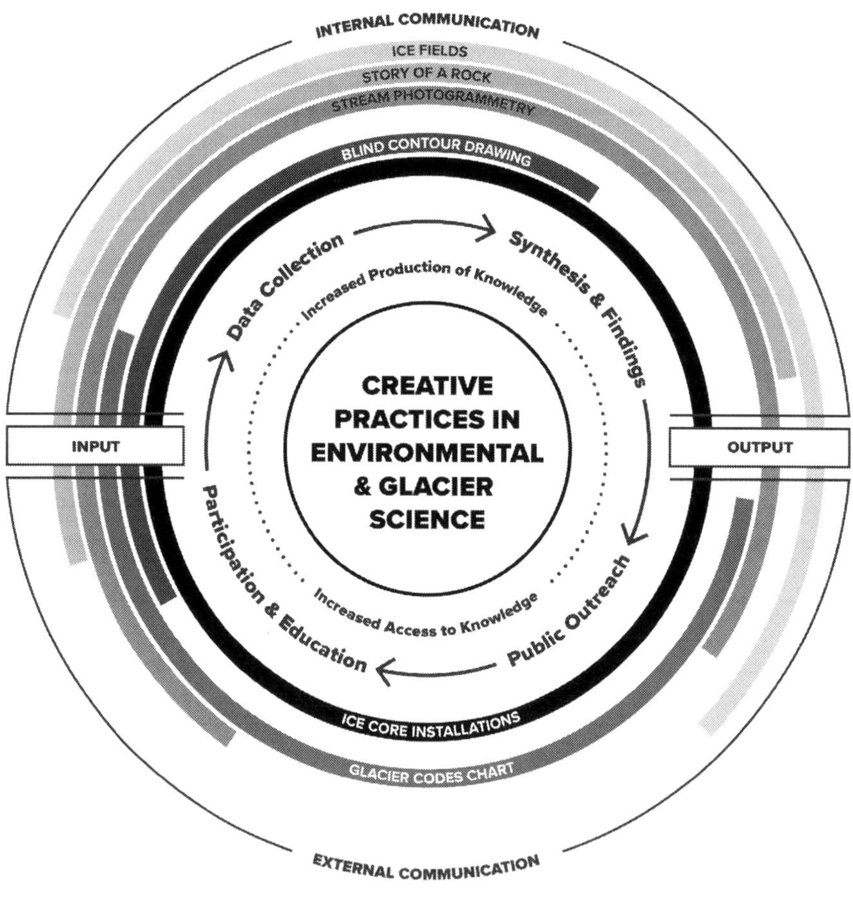

Figure 11.2 A proposed cyclical framework for integrating creative practices into scientific processes, featuring case studies mentioned in Section 11.5. Diagram by the author.

Figure 11.3 The project featured in the photographs are outlined in Sections 11.5.1 and 11.5.2. The numbered details below correspond to the numbered images in Figure 11.3.

(1) *Ice Fields*: The author's large format camera making a long exposure cyanotype print, overlooking the Taku Glacier and Camp 10, one of JIRP's permanent base camps. Photo by the author. (2) *Ice Fields*: A selection of cyanotype photographs of Camp 10 from summer 2023. Photo by the author. (3) *Story of a Rock*: A JIRP student working on writing and drawings to document their selected rock in 2022. Photo by Ben Huff. (4) *Story of a Rock*: Completed student research zines in 2023. Photo by the author. (5) *Stream Photogrammetry*: Students taking

(*Continued*)

Figure 11.3 (Continued)

measurements at the supraglacial stream on the Taku Glacier. Photo by the author. (6) *Stream Photogrammetry*: Students collaborating to stitch together printed cyanotype fabric of the supraglacial stream. Photo by the author. (7) *Glacier Code Chart*: Newly printed glacier codes with the historic "Cloud Code Chart," inspiration for this project. Photo by the author. (8) *Glacier Code Chart*: Finished and folded chart. Photo by the author. (9) *Ice Core Installations*: The author's MFA Thesis exhibition, *Planetarium*, at Mills College Art Museum in 2018, featuring hanging ice cores. Photo by the author. (10) *Ice Core Installations*: Viewers watch an ice core sculpture melt at OOBLECK Projects in San Francisco, CA, in 2017. Photo by the author. (11) *Blind Contour Drawing*: The author teaching the blind contour method above the Taku Glacier on the Juneau Icefield in 2022. Photo by Ben Huff. (12) *Blind Contour Drawing*: JIRP students practice the blind contour method overlooking the Vaughan Lewis Icefall in 2024. Photo by the author.

each fall within the framework diagram. It should be noted that the work of teaching and making art often blurs the lines between categories, methods, and subjects, and this fluidity is reflected here.

11.5.1 Internal: Data Collection and Synthesis

11.5.1.1 Art-Making: Ice Fields

Using a large-format camera, glacial meltwater, and the cyanotype printmaking process, I create long-exposure photographs of glaciers that document changes to the glacial environment. The visual outcome of each image is dependent on glacial albedo – how much light is reflected off the surrounding ice and snow – and captures entire days of light. These images become a record, not only of the surrounding environment but also of the daily human experience of fieldwork, as my ability to make a print is dependent on conditions and fieldwork schedules.

This body of work was included in the exhibition, *Extra Tough: Women of the North*, at the Anchorage Museum in 2021.

11.5.1.2 Teaching: Story of a Rock

The project, Story of a Rock, was developed in collaboration with Ben Huff and later Dr. Elizabeth Case and then taught in varying forms with many other participating faculty members. Each student creates a zine – in this case, a small handmade booklet – from a single sheet of paper. The workshop is divided into two parts. First, we have a place-based lesson about the local geology, including an introduction to the different types of rocks and guided discussions about how the local geology formed. Depending on the Icefield location, we explore various small intrusions, dykes, and fault lines. We also examined three types of igneous rock with different grain sizes, mineral compositions, and colors. We look, touch, and discuss different features of rocks and can recognize markings and directionality of where the glacier has flowed over them.

After the geology lesson, participants choose a rock to use as the basis for our observation project and zine. This selection is an intentional way to scale down the immensity of the landscape around us to a more specific and digestible focus for the day, but one that may be representative of larger, wider reaching concepts about the Icefield system and glacier geomorphology in general. We then lead a guided series of activities to help focus on each rock's texture, shape, and shadow. The goal is to teach basic techniques for documenting and sharing information about geology from various scientific and artistic perspectives. To do so, participants use an array of methods, like writing, field sketching, texture rubbings, scientific illustration, collage, and imaginative word-building.

11.5.1.3 Teaching: Stream Photogrammetry

In summer 2023, Dr. Elizabeth Case and I embarked on a new cross-disciplinary initiative, blending collaborative teaching, art-making, and research on the Icefield. Our aim was to explore creative methods for research, experimentation, data collection, synthesis, and communication. We led a multi-day series of art-science photogrammetry workshops to map and visualize an emergent supraglacial stream on the Taku Glacier. Photogrammetry is a technique that uses images to measure environmental features like size, scale, and color. By overlapping images and identifying reference points, it is possible to create detailed 2D images called orthomosaics, which help generate digital elevation models (DEMs) of the environment.

Over the course of four full-day or half-day workshops, students produced a mix of artistic and scientific materials: cyanotype prints of geologic and biologic matter, a 3D model of the field camp, a fabric tapestry depicting a nearby supraglacial stream system, and a DEM of this system for further analysis.

We first taught students how to operate remotely piloted aerial vehicles (RPAVs). They each made a geometric painting that could serve as ground control points for aerial surveys. We discussed color theory and design to ensure the images were as visible and accurate as possible. Students practiced flying the drone over our field camp, balancing factors like altitude and battery life as they spotted their strategically placed painted ground control points.

The second workshop focused on cyanotype printmaking, where students first learned about the history and methods of the cyanotype process and later experimented with the medium to make photographic "blueprints" of found materials like snow, ice, alpine plants, and rocks.

During the third workshop, we mapped the supraglacial lake system and studied stream dynamics. Students placed and surveyed ground control points, flew the drone over the system, and took measurements of channel depth and water velocity using various methods. At the stream, students improved their cyanotype skills and practiced safety techniques, using a rope system to safely lower themselves to dip cyanotype fabric into the glacial stream water to make abstract prints.

In the fourth workshop, we fully integrated cyanotype printmaking and photogrammetry concepts. Students used Metashape Pro to create a DEM from images of Camp 10, the drained lake, and the supraglacial stream.

We printed 12 overlapping negative images of the stream system onto transparency paper and then used them to create cyanotype prints of each image on fabric pieces. The students then stitched these images into a tapestry, literally finding their "tie points" and "stitching" the images together using embroidery thread. We proudly displayed this tapestry in the common kitchen for the program participants to observe throughout our time at that field camp.

11.5.2 External: Outreach + Participation

11.5.2.1 Art-Making: Glacier Code Chart

This classification system and text take inspiration from the Cloud Code Chart (1973), a system used to observe and record weather data. Cloud Code Chart was created and distributed by the U.S. Department of Commerce, the National Oceanic and Atmospheric Administration, and the National Weather Service. I first came across this chart at a JIRP field camp in 2017. The Glacier Code Chart is an original taxonomy and shorthand for recording glacial features, to aid in observation and understanding of the ice they carry and the surrounding land they transform. Much like clouds (bodies of water in other forms), glaciers are continuously in the process of change.

This chart is meant to prompt close looking and consistent observation as practice. In the same way that Cloud Code Chart is a reference and tool for learning, Glacier Code Chart is a useful teaching tool for glacial fieldwork. It was originally risograph-printed at Directangle Press in Bethlehem, NH, and many have been reprinted and brought back into the field on educational programs and other exhibitions. Featuring United States Geological Survey (USGS) definitions, designed symbols, and my own photographs of glaciers taken in Alaska from 2017 to 2022, these prints are educational both within and beyond a pure science context.

11.5.2.2 Art-Making: Ice Core Installations

Through scientific fieldwork and collaboration, I collect water samples, glacial silt, or moraine till and refreeze one or more of these elements into cylindrical molds to make ice sculptures. Each has cordage frozen into its center, which is often connected to a larger dynamic rope system – multiple pieces of ice attached to the same rope causes the installation to shift based on the rate of each ice core's melt. Dripping sounds are present throughout the space as the ice melt collects in/on ceramic, glass, or paper vessels below.

In each of these installations, I attempt to scale down geologic time and Earth's systems, like glacier melt, sea level rise, and erosion, to a human level in the gallery to make these concepts more accessible. In a gallery space, the sculpture needs consistent and repeated activation to continue to include ice. This refreezing process, which I call "Tending to the System," physically engages my (or a gallery attendant's) body into the sculptural process as well as the viewer's. In this way, as human influence is implicated in the depleting ice, so is attention, care, and stewardship.

11.5.2.3 Teaching: Blind Contour Drawing

Taking place across the Icefield, field sketching exercises cultivate strategies for recording scientific research and documenting experience. Topics range from lessons on perspective to contour drawing, scientific illustration, note-taking, and shading. While we teach a variety of field sketching methods, the most accessible and impactful is blind contour drawing.

The blind contour method is an exercise where the artist looks only at the subject, and not at their paper, drawing in one continuous line. This method was popularized by artist and educator Kimon Nicolaïdes in *The Natural Way to Draw* in 1941 and later reinforced by Betty Edwards in *Drawing on the Right Side of the Brain* in 1979.

The artist traces their eye around the "contours" of the subject very slowly, while attempting to match the speed and directionality of their sight path with their hand and drawing implement. This process is shown to enhance observation skills and develop hand-eye coordination [17], reduce perfectionism and anxiety [18], and promote holistic thinking [19].

Specifically on the Icefield, blind contour drawing workshops emphasize the value of drawing as a process, not just a product. As perfectionism is impossible, this practice creates vital space for experimentation, meditation, play, and deep observation. Blind Contour Methods help scientists at all levels cement features to memory and more fully understand complex visual information [20].

11.6 Conclusion

Addressing the global crisis of human-caused climate change requires interdisciplinary and collaborative approaches, supported by spaces and frameworks that facilitate such engagement. Creative practices play a crucial role in polar science education and research by amplifying knowledge production, diversifying learning experiences, and enhancing communication. Both within scientific communities and in public outreach, the integration of art can foster deeper personal connections, reduce political divisions, and create meaningful opportunities for reflection, empathy, and discourse. Embracing innovative and creative methods is essential for engaging both scientists and the broader public, promoting environmental stewardship, and advancing a more inclusive and holistic understanding of Earth science.

Addressing the pressing global crisis of human-caused climate change necessitates interdisciplinary and collaborative approaches that are supported by frameworks encouraging such engagement. Creative practices play a vital role in polar science education and research by amplifying knowledge production, diversifying learning experiences, and enhancing communication.

By fostering deeper personal connections and offering unique avenues for reflection, empathy, and discourse, the integration of art can bridge gaps in understanding complex climate science and glacier retreat, both within scientific communities and in broader public outreach.

Ultimately, my experiences, particularly through the JIRP, demonstrate the powerful impact of integrating creative methods throughout the scientific process, from data collection and synthesis to outreach and participation. This interdisciplinary approach not only enriches the study of rapidly changing glacial environments but also promotes environmental stewardship and a more inclusive and holistic understanding of Earth science by valuing diverse forms of knowledge production and learning styles. Embracing these innovative and creative entanglements is crucial for effectively engaging both scientists and the public in understanding and responding to the global climate crisis.

References

1. Kisner J. *Thin Places: Essays from In Between*. 1st ed. New York: Farrar, Straus and Giroux; 2020.
2. Vanni I, Crosby A. Edge. *Environ Humanit*. 2023;15(1):164–7. doi:10.1215/22011919-10216217
3. Gardner AS, et al. A reconciled estimate of glacier contributions to sea level rise: 2003 to 2009. *Science*. 2013;340(6134):852–7. doi:10.1126/science.1234532
4. Carey M, Jackson M, Antonello A, Rushing J. Glaciers, gender, and science: A feminist glaciology framework for global environmental change research. *Prog Hum Geogr*. 2016;40(6):770–93. doi:10.1177/0309132515623368
5. Tosca MG, Galvin A, Gilbert I, Walls KL, Tyler GE, Nastan AM. Reimagining futures: Collaborations between artists, designers, and scientists as a roadmap to help solve the climate crisis. *Elementa Sci Anthropocene*. 2021;9(1):00016. doi:10.1525/elementa.2021.00016
6. Ivanitskaya L, Clark D, Montgomery G, et al. Interdisciplinary learning: Process and outcomes. *Innov High Educ*. 2002;27(2):95–111. doi:10.1023/A:1021105309984
7. Kaikkonen L, Shellock RJ, Selim SA, et al. Fostering diversity, equity, and inclusion in interdisciplinary marine science. *NPJ Ocean Sustain*. 2024;3:49. doi:10.1038/s44183-024-00087-1
8. Ripple W, Newsome T, Barnard P, Moomaw W, Macé M. World scientists' warning of a climate emergency. *BioScience*. 2020;70(1):8–12.
9. Bruine de Bruin W, Rabinovich L, Weber K, et al. Public understanding of climate change terminology. *Clim Change*. 2021;167(37). doi:10.1007/s10584-021-03183-0
10. Gorsegner A. Role of art in the global climate change movement [doctoral dissertation]. Drexel University; 2016. doi:10.17918/etd-6754
11. Li Q, Marshall J, Rye CD, Romanou A, Rind D, Kelley M. Global climate impacts of Greenland and Antarctic meltwater: A comparative study. *J Climate*. 2023;36:3571–90. doi:10.1175/JCLI-D-22-0433.1
12. Haim K, Aschauer W, Weber C. The effectiveness of the teaching program "scientific creativity in practice". *Discov Educ*. 2024;3:255. doi:10.1007/s44217-024-00368-4
13. Khoury CK, Kisel Y, Kantar M, et al. Science–graphic art partnerships to increase research impact. *Commun Biol*. 2019;2:295. doi:10.1038/s42003-019-0516-1
14. Keener C, Holzman J, Rigor I, Woods J. Arctic Ice: A Visual Archive: A unique collaboration among Cy Keener, Justine Holzman, Ignatius Rigor, and John Woods. Washington, DC: The National Academies Press; 2023. https://nap.nationalacademies.org/catalog/26948/arctic-ice-a-visual-archive-a-unique-collaboration-among-cy

15. Lan X, Wang Y, Peng L, Ma X. More Than Beautiful: Exploring Design Features, Practical Perspectives, and Implications of Artistic Data Visualization. 2025. doi: 10.48550/arXiv.2502.04940
16. Hughes BS, Corrigan MW, Grove D, et al. Integrating arts with STEM and leading with STEAM to increase science learning with equity for emerging bilingual learners in the United States. *IJ STEM Ed*. 2022;9:58. doi:10.1186/s40594-022-00375-7
17. Tchalenko J. Segmentation and accuracy in copying and drawing: Experts and beginners. *Vision Res*. 2009;49(8):791–800. doi:10.1016/j.visres.2009.02.012
18. Drake JE, Winner E. How children use drawing to regulate their emotions. *Cogn Emot*. 2013;27(3):512–20. doi:10.1080/02699931.2012.720567. Epub 2012 Sep 11. PMID: 22963448.
19. Edwards B. Drawing on the right side of the brain: A course in enhancing creativity and artistic confidence. Los Angeles: J. P. Tarcher; 1979.
20. Fan JE, Bainbridge WA, Chamberlain R, et al. Drawing as a versatile cognitive tool. *Nat Rev Psychol*. 2023;2:556–68. doi:10.1038/s44159-023-00212-w

12 Arctic Encounters

Material Culture, Indigenous Worldviews, and AI Projections

Elisa Palomino and Jonathan Katz

12.1 Introduction

12.1.1 Arctic Fashion: Design, Environment, and Identity

Fashion is a cultural, material, and economic phenomenon expressed through dress and body adornment. It is a global practice that extends to remote Arctic regions [1]. The study of Arctic fashion showcases how clothing reflects both environmental adaptation and cultural identity, revealing material resources, ecological changes, and human ingenuity. The Arctic's climate demands clothing that prioritizes protection and functionality. Nowhere else does nature exert such a significant influence on clothing design. Arctic skin garments represent some of the most advanced adaptations to extreme cold, crafted from a variety of animal skins, including those of mammals, birds, and fish [2]. Despite shared environmental challenges, regional differences emerge in the materials used, construction techniques, and styles of clothing. These variations reflect localized responses to the natural environment, as well as cultural distinctions among Arctic Indigenous Peoples. Garments are chosen according to seasonal needs, and specific sewing techniques or embellishments containing knowledge systems embedded in their construction often signify identity markers such as gender, age, status, or ethnicity. This dual function of clothing—to protect the body and communicate cultural identity—makes it a powerful subject of anthropological inquiry. Clothing in the Arctic is more than a means of survival; it is a powerful medium of cultural expression. It serves as a non-verbal communicator of values, traditions, and identity, asserting cultural distinctiveness and reaffirming a sense of belonging within communities [3].

Arctic fashion also responds to environmental and societal shifts. Biodiversity loss impacts the availability of traditional materials, prompting innovations that blend Indigenous knowledge with contemporary design practices. This evolving relationship between Arctic clothing and the environment reflects the resilience and adaptability of its makers. Moreover, Arctic fashion sustains cultural knowledge, preserving skills and traditions that link people to the land and to one another [4]. Arctic fashion exemplifies a sustainable and reciprocal relationship with the environment. It is rooted in local ecosystems, drawing on

DOI: 10.4324/9781003545880-16

materials that the land provides while ensuring minimal waste. The creation of garments under conditions of material availability highlights the ingenuity of Arctic communities in transforming natural resources into objects of cultural and functional significance [5].

As an artistic practice, Arctic fashion bridges the material and symbolic, the functional and the expressive. It reflects a profound connection to the environment while serving as a means of cultural preservation and identity affirmation. By exploring regional variations, environmental adaptations, and the social dimensions of Arctic clothing, we gain an understanding of how material culture can address contemporary challenges, including sustainability, Indigenous rights, and the impacts of biodiversity loss.

12.1.2 Arctic Indigenous Peoples

Arctic Indigenous Peoples belong to diverse regions, with distinct nationalities and unique cultures, yet they share features of harsh climates, natural environments, dispersed populations, and ancient practices [6]. Their Indigenous knowledge systems and technologies around engaging more effectively with nature represent highly innovative contributions to the global future. These range from sustainable fishing, hunting, and processing of organic materials to respectful animal husbandry and a mutually beneficial relationship with the surrounding environment. They have skillfully performed these sustainable processes for millennia [7].

12.1.3 Colonialism in Anthropology Museums

For centuries, anthropology museums have played a crucial role in expanding the general public's cultural awareness and understanding of human and cultural diversity. However, in the past two decades, museums have evolved into spaces marked by tension and conflict [8]. Internationally, museums have curated ethnographic collections featuring ritual objects, clothing crafted from animal skins, tools, and various household items. Often these collections were assembled to document the material culture of what was wrongly assumed to be disappearing cultures [7].

Colonialism significantly impacted Indigenous Peoples' cultures worldwide, including Arctic and subarctic regions. Explorers, missionaries, and collectors removed material culture from communities throughout the Arctic, relocating them to museums in non-Arctic regions, creating a disconnect between Natives and their material heritage. While curators preserved knowledge about these artifacts, the link to their original communities was frequently severed. This detachment was further intensified by the practice of inter-museum trade, which aimed to provide a comprehensive representation of global cultures. This led to the dispersion of pieces and loss of knowledge. Local communities found themselves unaware of the material heritage of their ancestors, and museums lost crucial insights into the intangible heritage associated with the collection pieces they were safeguarding [9]. When communities realized that their ancestral heritage must be returned, the key concept of access to collections emerged [10] and institutions assumed their responsibility to disseminate knowledge about the Indigenous artifacts in their collections. Geographical distance and prohibitive travel costs further hinder direct community access to these dispersed collections. To address these challenges, institutions organize visits of Elders, traveling from remote Arctic areas to spend periods with the collections. However, this method limits the participation of Arctic and subarctic communities in these experiences.

Many Galleries, Libraries, Archives, and Museums (known collectively as GLAMs) contribute significantly to international tourism, research funding, creative industries, and economic drive. Meanwhile, their associated Indigenous communities lack access to their tangible and digital cultural heritage, or the resources to establish their own cultural institutions and digital collections. Recently, calls to decolonize GLAMs have come from the Indigenous communities dispossessed of their cultural heritage [11]. Their continued marginalization has led to significant representational gaps in global culture, as their achievements have been systematically erased and undervalued. Sharing these resources can help secure these lost testimonies, thus repairing historical legacies [12]. Decolonization begins with acknowledging the systemic inequalities propagated by colonialism and wealth extraction, revealing how heritage was obtained in the first place by making collections more representative of communities' histories [11].

12.1.4 Museums, 3D Technologies, and Indigenous Engagement

The synergy between museums and AI-enhanced virtual spaces is reshaping the future of cultural institutions, facilitating experimentation with advanced technologies through collaborations between Indigenous Peoples, researchers, and curators. The use of 3D technologies in studying Indigenous material culture accelerates research methodologies, offering visualizations and prototypes while providing new ways to experience arts and culture online. 3D-generated artwork and interactive images enable the creation of immersive assets as storytellers and emotion evokers. Museums, traditionally underutilized, can take advantage of digital tools to connect their collections and expand access to Intangible Cultural Heritage (ICH), transforming the way we experience heritage. Anthropology museums could serve as a model, integrating digital tools to engage with the community for meaningful connections to collections while decolonizing their management.

The use of digital 3D models and replicas in caring for Indigenous collections, including digital repatriation, connects museums with origin communities, complementing physical repatriation efforts [13, 14]. While digital repatriation extends beyond returning physical objects, it opens avenues for increased community access and contributes to cultural heritage preservation [9]. Alongside this, associated materials, digital material, and intellectual property rights should be part of the repatriation process. In addition, the application of these technologies to historical Indigenous artifacts raises ethical issues. In the past, the categorization of artifacts in institutions was simplified by the joint classification of many objects, which reduced the complexity of Indigenous knowledge. In addition, some of the data in the collections have a racist legacy and contain information detrimental to Indigenous cultures; this content is not suitable for digitization by 3D modeling [15]. The intersection of museums and digital humanities is pushing the heritage sector to innovate while grappling with many ethical challenges.

12.1.5 Anchorage Museum at Rasmuson Center

The case study below showcases the digital replication of a 19th-century Yup'ik fish skin parka at the Anchorage Museum in Alaska. Founded in 1968 by the Rasmuson family, it has become Alaska's largest museum and features an art collection and an Alaska First Peoples Gallery with 600 objects on loan from the Smithsonian. The Arctic Studies Center, part of the Smithsonian Institution's National Museum of Natural History, has been a partner of the Anchorage Museum since 1994, actively working with Alaska Native Elders for two decades, facilitating the return of their ancestors' artifacts, such as fish skin containers and fish skin parkas. The joint effort resulted in the exhibition "Living Our Cultures, Sharing Our Heritage: Alaska's First Peoples" at the Anchorage Museum [16]. This exhibit showcases Alaska Native historical artifacts and highlights community involvement in hands-on study in the places of origin of the artifacts. The inclusion of a cultural resource center, where Indigenous Elders teach traditional skills, a study center, and an archaeology lab underscores the commitment to a community-centered approach to presenting Alaska Native history, aligns with decolonization efforts, and contributes to inclusivity in museum representation of Indigenous cultures.

12.1.6 3D Modeling

3D modeling involves using computational design tools such as Blender or CLO3D. An example is the case study below of the 3D recreation of a fish skin parka, where human expertise drives the design process. Although AI tools generate imagery according to general visual resources, the core of 3D modeling remains a human mind-led activity.

Digitalization is the broader concept under which 3D modeling falls. It refers to the comprehensive process of converting physical objects, knowledge, and cultural heritage into digital formats. Within this framework, 3D modeling serves as a key tool for digitizing physical artifacts, enabling the preservation and study of cultural heritage in virtual form. These technologies fulfill distinct yet complementary roles, with digitalization serving as the unifying concept that integrates their contributions.

In our work, we have leveraged sculpting and modeling tools to enhance the creation of 3D fish skin texture. Tools like Blender and CLO3D incorporate these features as part of their functionality. These features allow automatic software tools while still allowing the use of analogue and organic design methods.

12.1.6.1 3D Modeling: Replicating a 19th-Century Yup'ik Fish Skin Parka

The original Yup'ik fish skin parka, housed at the Anchorage Museum, was part of Elisa Palomino's research during her 2019 Fulbright Fellowship at the Arctic Studies Center. Monica Shah, Assistant Director of Conservation and Collections, introduced Elisa to the museum's collection of fish skin objects. Elisa, in turn, invited June Pardue, an Alutiiq and Inupiaq elder renowned for her expertise in native arts, to examine the parka and share her knowledge of materials and sewing [17]. With a diverse artistic repertoire that includes weaving grass, sewing fish and marine mammal skins, storytelling, and beadwork, June Pardue is an Adjunct Professor of Native Arts at University of Alaska, Anchorage; she is a regular lecturer at the Anchorage Museum's cultural resource center, the Alutiiq Museum, the Hermitage Museum in St. Petersburg, and the Autry National Center in LA.

During the analysis of the Yup'ik fish skin parka, we discussed its formal design elements, construction, and the spiritual and economic significance of fish skin. June Pardue's invaluable firsthand knowledge enriched the interpretation, shedding light on the materiality and cultural heritage associated with fish skin. The focus extended to capturing the memories embedded in the garment through details like the choice of fish species, tanning and sewing technology, wear patterns, and repairs. In this collaborative exploration, June Pardue, through storytelling and sharing of ancestral knowledge, played a crucial role in attributing cultural value to the fish skin artifact and deciphering its broader cultural significance.

Alaska Natives used fish skin to make clothing for practical and ceremonial use. Women sewed parkas for adults and children, as well as hunting clothing for men. The fish skin material and clothing are part of a belief system; hence, a fisherman would wear a fish skin parka for protection when fishing for salmon [18]. The most commonly used fish species were king and chum salmon. Female king salmon provided the best material, as their skins were larger and thicker. Chum salmon skins were valued for their durability, and both were lightweight. Smaller sized pike skins were used for children's clothing. Women also used halibut, turbot, and eel skins. A parka was always made from the local fauna of the region [19].

This is a small pike-skin parka or *qaspeq* (thin hooded garment) destined for a child and has rounded edges trimmed with a calfskin fringe. The main core of the parka is made of white fish skin, with the inside of the skin placed on the outside. On the lower part of the front of the parka are two black inserts called *uminguak* (two arrowhead designs), reminiscent of the times of bow and arrow warfare [20]. The parka pattern is created by manipulating the subtle shades of fish skins [21]. Earth tones were used to dye the parka, which were the most common in Arctic societies, which had access only to natural raw materials [22]. Mineral dyes such as red ochre (hematite) were used to provide tonal contrasts in leather decoration [23]. It could also be soaked in fish blood and placed in the fire to redden [24]. It was used wet on the end of a stick and applied to the seams of bags, boots, and parkas to add strength while providing embellishment [25]. Red ochre was a valuable commodity [26] and considered to have supernatural powers. The places where it was found were considered sacred and offerings were left when it was taken [27]. It was one of the strongest colors available and was used to dye skins while acting as a preservative. In this parka, the pieces of fish skin around the hood and shoulders have been dyed red to add strength to the seams. Carbon black was made from charcoal mixed with fish roe (caviar) and used to produce black tones. Bleaching was achieved by Alaska Natives using solutions of stale urine with a high ammonia content [21]. Another method of bleaching was to hang untreated skins in the frost and leave them in the open air until they turned white [28]. Arctic seamstresses used these contrasting colors and the varied shapes of the skin to create beautiful patterns.

The parka has been sewn with the same waterproof stitch used when sewing seal gut, including coarse seagrass as reinforcement on both sides of the seams. Often used in combination, the two could swell with moisture and keep the parka waterproof [29]. A regular stitch or an overcast stitch could also be used on the fish skin. The sinew could serve as thread, but some made thread from the fish skin itself. In fact, there is archaeological evidence of overlock stitches, running stitches, and waterproof stitches [30]. Moreover, fish skin parkas were lightweight and had the ability to repel water and block wind [20].

In the unwritten society of Alaska Natives, the designs and symbols on a parka are a way of conveying important information, connecting with each other and the environment. A parka acts like a book that tells a story; it is a beautiful art form with a wealth of material and information about a family history. A parka is also like a house, protecting a person from the harsh Arctic environment. The learning process for the successful procurement and preparation of animals to make a parka begins in early childhood. Eventually, all forms of life come together through the hands and skill of a seamstress [19]. The skills of Arctic seamstresses have been praised by explorers throughout history [31]. Custom-made clothing, tailored to fit a particular person's dimensions, was essential for survival in extreme Polar environments [32]. To make waterproof and durable clothing, women had to design a technology that could keep the garments together and dry [33]. During colonization, explorers faced extreme weather conditions, and a well-stitched fish skin parka could be a lifesaver. Collaboration with Indigenous Arctic Peoples, who shared knowledge of the local climate and materials technology, played a crucial role in the explorers' survival. Subsequently, Arctic inventions and technologies, such as parkas, were adopted around the world. Indigenous Arctic Peoples deserve credit for the existence of this outerwear, and major outdoor clothing companies such as Patagonia and North Face have been inspired by the ingenious designs of Arctic seamstresses, a gift shared with the world [34].

12.1.6.2 3D Recreation of a Yup'ik Fish Skin Parka by Jonathan Katz

Jonathan Katz, an MA illustration student at the Royal College of Art in London, used a 3D software to digitally reproduce the Anchorage Museum's 19th-century Yup'ik fish skin parka. In collaboration with Elisa Palomino, the project engaged communities in Washington, Anchorage, and London, offering new avenues for participation in heritage activities. The study investigated the historical significance of Arctic clothing, with an emphasis on the role of Indigenous Arctic Peoples in influencing global outerwear. The process involved digitizing the parka using computational design tools. The software used in the study was examined for its integrated nature, used creatively in the design process.

Details of the fish skin parka were photographed with a digital camera (Figure 12.1). Photographic samples were collected to reconstruct the texture of the fish skin. Arctic and subarctic ethnographic collections often consist of organic and soft materials, which are difficult to model in 3D, as the pieces are a composite of several materials. It is crucial to depict soft materials, such as fish skin, and hard, shiny surfaces, such as glass or ivory beads. Generally, 2D images omit significant details for members of the original community, while 3D digital models offer advantages, as they allow users to rotate and view the artifacts, experience the changes in light, and, at the same time, interact remotely with the cultural heritage [35].

Jonathan used the shape of a salmon skin to create a module, then introduced parametric design tools to construct the parka. In order to preserve the aesthetic qualities of the fish skin garments, he reconstructed the iconography from its minimal graphic unit: the texture of the fish skin. The virtual reality

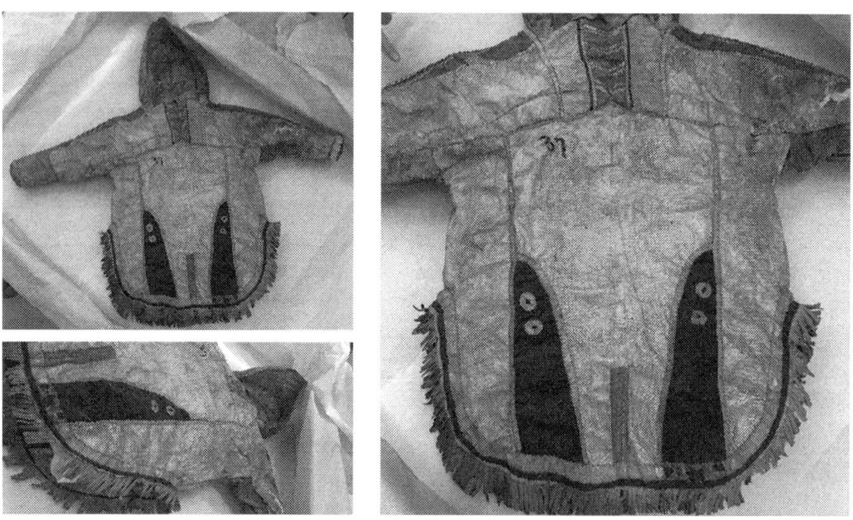

Figure 12.1 Digital photography of the Yup'ik pike skin child's parka. c. 1900. Anchorage Museum.

replica of the Yup'ik parka was made using two software packages: CLO3D [36] and Blender [37]. CLO3D, a state-of-the-art three-dimensional software that combines pattern making with full-featured 3D virtual simulations, is easy and fast to use [38]. Blender can be used to recreate a simulation of materials and rigid bodies and to model the shell of the artifact [39]. A virtual construction model is created based on augmented reality technology used in the garment manufacturing process to simulate the cut and drape of the fabric according to the simulated parameters of the fabric attributes.

Jonathan drew a pattern block of the parka on the contours of the photographs (Figure 12.2). The pattern block was then exported to CLO3D using the program's functions to reconstruct the garment pieces in 3D and virtually

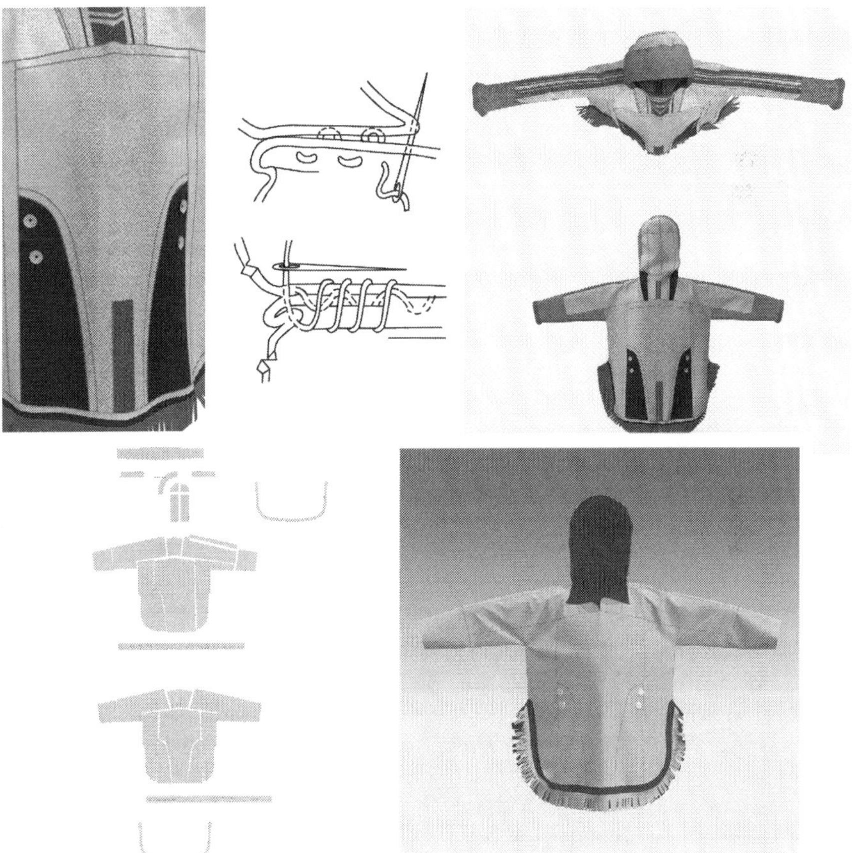

Figure 12.2 Clockwise from top left: (Top left) Waterproof stitches used to sew fish skin garments. Illustration by Jonathan Katz. (Second from left) Waterproof stitch. Illustration by Jonathan Katz. (Top right) Top view and back view of the 3D modeled fish skin parka. Illustration by Jonathan Katz. (Bottom right and left) Parametric tools used for the integration of fish skins in the creation of the fish skin parka. Illustration by Jonathan Katz.

cut them. The "Sew" command was used to sew the pieces together. Physical properties such as the elasticity of the fish skin, its density, and thickness are some of the functionalities enabled by the program. The properties of pike fish skin were chosen for this children's parka (Figure 12.1). The default leather parameters from the CLO3D library were used and fish scales were added, creating a new parameter. The different shades of the fish skins were manipulated to closely resemble those of the original parka. The skin drape prediction regression model was used to create the simulated material using attribute parameters [40], creating a three-dimensional volume by stitching triangular fish skins (Figure 12.2).

Certain sewing techniques mastered for centuries by Arctic seamstresses, such as the waterproof stitch (Figure 12.2), do not exist in CLO3D's sewing repertoire, so by including this unique technique, the program is acquiring new knowledge. The use of 3D software should serve not only aesthetic purposes but also functional ones. Recreating a garment does not merely serve to create a virtual object, but by collaborating with the communities, we reconstruct it to provide virtual access and enrich it. Moreover, the digital representation of Indigenous crafts, especially within software like CLO3D, introduces additional complexities such as the software's reliance on predominantly "western" sewing knowledge, potentially leading to difficulties in faithfully representing the intricacies of Indigenous craft techniques. This challenge not only raises questions about the inclusivity of digital platforms but also highlights the importance of addressing cultural nuances and diversifying the knowledge bases embedded in such technologies.

When exploring the impact of new AI regulations on international cooperation, particularly in the realm of heritage preservation and the intersection of digital technologies, it is essential to be mindful of potential biases and limitations. One concern is the risk of perpetuating biases in racial representation, as digital platforms and design software may predominantly generate representations aligning with white European beauty standards. The unintentional propagation of such biases poses a challenge in the realm of heritage preservation, where the authenticity and diversity of cultural artifacts, peoples, and practices must be accurately captured and preserved.

In Alaska Native collections, there is often little information about the artist, the community of origin, or the materials used. These digital technologies provide opportunities for craftspeople to obtain information about traditional techniques from culturally significant objects made by ancestors, enriching the understanding of these objects. 3D technology facilitates the digital reunification of communities with their heritage, helping to decolonize collecting practices by eliminating the need to physically remove objects for documentation. Experimentation in museum practices, combining digital tools with analogue activities, is crucial to developing more sustainable practices. This research lays the foundation for visual analysis of Indigenous materials and is scalable to any artifact, offering an opportunity for the dissemination of Arctic cultural heritage.

The combination of different technologies and digital applications can be attractive to Indigenous youth seeking to combine craft techniques with 3D modeling to facilitate cultural learning through playful discovery. Digital prototyping of museum artifacts does not replace traditional material knowledge; instead, they support each other, enabling innovation. The digital environment allows the animation of historical garments in a new context, maintaining accuracy while enabling material, shape, and texture manipulation. The intersection of advanced digital technologies, sustainability, and contemporary museum practices emphasizes the importance of connecting anthropology, ethnography, and craftsmanship [41]. This project laid the groundwork for the visual analysis of Indigenous materials. The use of 3D technology contributed to repatriating cultural heritage, sharing information about artifacts, and decolonizing collection practices.

12.1.7 Next Steps on Digital Repatriation

During the presentation of the preliminary results of this project at the EFHA European Fashion Heritage Association Conference in Utrecht, we were contacted by the marketing manager of CLO3D, who expressed her interest in the project and the possibility of collaborating in the future.

In 2019, Elisa Palomino met with many Alaska Native fish skin artists. One of the pieces that attracted significant interest was the Alutiiq fish skin hood, which was collected from Alaska by Reverend Sheldon Jackson during the 19th century and is now preserved at the NMNH. All the artists expressed interest in the possibility of exhibiting the Alutiiq hood in its place of origin, Kodiak Island, under the roof of the Alutiiq Museum. During her travels, Elisa also saw a replica of the fish skin hood at the Alaska Native Heritage Center, but the accuracy of the measurements, pattern, and materials was not matched, probably due to the absence of the physically available artifact.

As the Alutiiq Museum was under renovation at the time of this work, we considered organizing a visit to NMNH with Alutiiq Elders, who will travel to Washington to research the original piece, photograph and videotape it, recall how the object was made and used in the past, and facilitate the creation of new knowledge about the artifact. Once back on Kodiak Island, a CLO3D workshop will be organized with a group of local youth. CLO3D will provide the software and 3D modeling training to a member of the Alutiiq community, who will in turn train local creative community members. With the data collected by the Elders at the NMNH, a replica of the Alutiiq fish skin hood will be created.

Once the fish skin hood has been digitally repatriated, the physical repatriation of the real hood will take place and the NMNH will lend the piece to the Alutiiq Museum. A fish skin tanning and sewing workshop will be organized in which locals can participate. A small exhibition will showcase the original 19th century fish skin hood, 3D replicas made by Alutiiq youth, and physical replicas made by community members, bringing together past and present while contributing to the repatriation of cultural heritage.

12.1.8 Policies Regarding the Reproduction of Arctic Artifacts in 3D Environments

In our constantly evolving world, 3D modeling stands on the brink of revolutionizing museum practices. Creating a 3D prototype offers invaluable insights into museum artifacts by enabling remote visualization of these items. By breaking down barriers, 3D modeling is making museums more inclusive and accessible than ever before. Meanwhile, the impact of digitalization on the creative realm is a subject of intense debate, with some viewing it as a powerful tool enhancing human creativity and others seeing it as a challenge to creativity itself.

While the sharing of material culture creates opportunities, it can also create inequality and can lead to both benefits and harms [12]. Legal provisions have not been contemporaneously created to address the 3D technologies, necessitating museums to proactively prepare for this emerging digital space. It is crucial to delve into past legal precedents to ascertain how current contractual obligations can be extended to encompass existing Arctic artifact collections when digitizing them on emerging 3D platforms. Additionally, it is essential to contemplate the future implications for intellectual property and rights as the process of 3D asset digitization continues to evolve. This proactive approach ensures that museums navigate the legal landscape of 3D virtual platforms effectively and address potential complexities related to their digitized collections.

Policymaking plays a crucial role in addressing these challenges. Recognizing the increasing influence of digitalization on economic, social, political, and cultural aspects, a regulatory framework must be adopted to safeguard fundamental rights, address digital inequality, and foster technological innovation to ensure a responsible and ethical integration of digitalization into society. The extent of its impact on modern life is still being fully comprehended, prompting academics, professionals, policymakers, and legislators to scrutinize its effects on potential human rights violations and associated risks [42].

12.1.9 Indigenous Involvement in Shaping the Use of 3D Modeling in the Arctic

The UNESCO [43] Recommendation on the Ethics of Artificial Intelligence acknowledges the transformative potential of AI in addressing global challenges but warns against risks, urging inclusive participation in AI development. Emphasis is placed on the inclusion of Indigenous communities, recognizing their centuries-old knowledge, such as on climate change mitigation. Digitizing Indigenous knowledge using 3D modeling can aid in preservation and transmission, fostering cultural heritage. The need to democratize 3D modeling development, aligning with UNESCO's ethics that stress diversity and inclusion, is paramount.

While recommendations set guidelines for developing 3D modeling with Indigenous Peoples, it's crucial to address the diverse experiences among the more than 5000 groups globally. Overcoming challenges requires a focus on inclusion, decolonization, and local ownership. Collaborative efforts are essential, involving communities, researchers, and stakeholders in such development. This ensures alignment with Indigenous values, needs, and priorities, promoting a tool that respects human rights while addressing issues of culturally appropriate access.

The potential of 3D modeling to preserve cultural heritage requires a shift toward initiatives led by Indigenous voices in the decision-making process. The process should not be led by the cultures historically responsible for colonization, heritage, and wealth extraction, but by Indigenous Peoples [15]. The UN Declaration on the Rights of Indigenous Peoples [44] reaffirms their rights to self-governance and authority to control their cultural heritage. Therefore, policies must be community-designed for genuine representation. Indigenous-Centered 3D design must align with their perspectives, guarding against misuse and reinforcing technology with cultural competencies embracing and respecting the rich human experiences of Indigenous communities through ongoing collaboration and co-creation [45].

12.1.10 The Use of 3D Modeling and Indigenous Knowledge

Arctic Indigenous Peoples, deeply connected to spiritual traditions and intellectual wealth, embody a distinctive human culture intertwined with nature's co-evolution. Within Indigenous knowledge, there is knowledge specific to individuals, as well as knowledge related to specific clans or seasonal periods of the year [46]. However, globalization and environmental impacts threaten their traditional way of life. In the past decade, global attention to long-term digital heritage preservation led UNESCO to adopt the "Charter on the Preservation of Digital Heritage" and related guidelines in 2003. Digital cultural heritage encompasses digitized artifacts recreated in digital form, aiming for long-term availability while maintaining source material characteristics [6].

The Indigenous Artificial Intelligence Protocol, introduced in 2020 by the homonymous Working Group, offers a framework for designing AI systems, highlighting the need to embrace the multiplicity of Indigenous knowledge systems [47]. Recognizing the cultural essence of computational technology, 3D modeling designers should consider industry-dominant cultural frameworks, normative ideals, and contextual biases. Addressing biases inherent in culture, strategies should be devised to broaden cultural and social contexts in technology development [45].

12.1.10.1 3D Modeling and Cultural Appropriation: Navigating Complexities

Digitization can reignite and exacerbate issues around material culture collections related to ownership, authorship, repatriation, knowledge production, cultural memory, and open access. Museums must recognize the history of collecting, colonial pursuits, looting, and other impacts on the contemporary location of an artifact, its provenance, and the knowledge generated around it [11]. The ethical considerations surrounding the ownership, archiving, access, and usage rights of 3D models lack definitive guidelines and regulations. When dealing with Indigenous material culture, negotiations must carefully address cultural property rights and the rights of the model creator. Clear protection measures, collaboratively decided upon, are crucial to align digitized items with the cultural needs of the origin communities [48]. Archiving 3D models in public digital repositories with unrestricted access may not be suitable for certain Indigenous material heritage. Communities might opt to restrict access to comply with culturally specific limitations on sensitive artifacts, deeming items like medicine bundles or funerary objects as untouchable due to the presence of dangerous spirits [9].

As this tool evolves, incorporating Indigenous data sovereignty perspectives from the design stage becomes imperative to prevent reinforcing cultural appropriation conditions. The growth of 3D modeling brings risks such as the lack of attribution to authors' work [49] and unauthorized data use, perpetuating historical power inequalities that render Indigenous Peoples invisible and strip them of control over their cultural production. In recent years, Indigenous groups have invested substantially in protecting their intellectual property (IP) in the physical realm. The move into 3D virtual platforms should extend these efforts to safeguard the same IP in the virtual realm.

12.1.10.2 Recommendations for the Use of 3D Modeling on Indigenous Artifacts

Western concepts and values do not adequately protect Indigenous cultural expressions, neither reflect their cultural values which may include sacred or sensitive content. Ethical considerations must include how Indigenous knowledge is created, collectively maintained, and transmitted across generations following their customary laws and protocols [50]. Collaborative best practices informed by Indigenous input in the Arctic are essential and guidelines explaining 3D technology and policies aligning with Arctic needs should be established. Culturally responsible data sharing practices, developed in collaboration with local communities, can aid in preserving tangible and intangible heritage. The responsible use of 3D modeling technology and Indigenous data sovereignty requires participatory efforts, respectful communication, non-hegemonic perspectives, and leadership from the communities in technology development.

To approach this scenario, we offer the following recommendations:

12.1.10.3 Methodology

12.1.10.3.1 INDIGENOUS SOVEREIGNTY

- Design methodologies to avoid invasive relationships between academics and communities.
- Identify the sovereignty of the different Arctic Indigenous groups involved in the project.
- Research thoroughly the history of each Native group prior to each project.

- Acknowledge the original land the project is taking place upon, respecting the traditional owners.
- Establish a working group composed of the Indigenous artists involved.
- Hold meetings with Indigenous experts to broaden understanding of Indigenous knowledge.
- Explore ways to protect knowledge and material culture.
- Earn the trust of the community before Indigenous knowledge is shared.
- Use traditional ways of knowing and learning when passing on cultural knowledge to third parties.

12.1.10.3.2 CO-DESIGN

- Provide 3D modeling training to the community members as part of all projects.
- Provide software design accessible to users with varying abilities to maximize usability.
- Generate spaces for open innovation and technology building by Indigenous communities.
- Promote Indigenous Peoples' access to new technologies for the revitalization of heritage.
- Create mechanisms for collaboration in 3D modeling projects with a focus on Indigenous Peoples.
- Address biases on racial representation to broaden cultural contexts in technology development.
- Highlight the role of young Indigenous Peoples and new technologies, supporting these projects.
- Promote the development of Indigenous Peoples' 3D modeling technologies through grants.
- Establish dialogue between technology development companies and Indigenous Peoples.
- Facilitate the incorporation of Indigenous Peoples in the 3D modeling workforce.
- Generate culturally diverse learning spaces for technology developers of 3D modeling systems.

12.1.10.3.3 PROTECTION OF INTELLECTUAL PROPERTY RIGHTS

- Develop an understanding of the local customary laws that govern access to the techniques associated with Indigenous Cultural Heritage.
- Request Indigenous Peoples' right to give or withhold their "free, prior and informed consent" (FPIC) in any project involving intellectual resources that affect their cultures.
- Consult with Native artists on protecting their Indigenous knowledge from being shared with others.
- Provide regulations of ownership, archiving, access, and use-rights of 3D models.

- Carefully negotiate cultural property rights and the rights of the person creating the 3D model.
- Consult Native communities about what can be digitized and made available online.
- Recognize that archiving in public digital repositories that provide free access to all end-users is not an appropriate option for digital 3D models of Indigenous material heritage.
- Provide an accurate representation of Indigenous names and places for future generations.
- Identify items that are spiritually significant to an Indigenous community, and whose usage may be offensive, avoiding their use.
- Restrict access to certain 3D models to comply with specific restrictions of culturally sensitive artifacts in museums.
- Negotiate the use of 3D models with the tradition holders from the original culture.
- Identify individual versus collective rights in the context of Native laws and legislation.
- Consider and respect the wishes of the communities to avoid misappropriation of cultural heritage, constantly negotiating these relations.
- Consider whose cultural agendas are advanced by open access and the public domain.

12.1.10.3.4 COMPENSATION

- Create a written agreement clearly defining the responsibilities and rights of each party.
- Indigenous Peoples should be fairly compensated, which could include economic compensation, acknowledgment as author, co-author or contributor, royalties, copyright, patent, trademark, or other forms of compensation.
- If there are economic gains involved, the end results should benefit the Indigenous Peoples.
- Provide, in return, additional sources of knowledge and expertise.

12.1.10.3.5 COMMUNICATION OF RESULTS

- Provide periodic updates to share progress and to ensure that people feel informed.
- Provide recognition of all Indigenous artists and their work.

12.1.11 Conclusion

The chapter has explored the impact of new AI regulations on international cooperation and transnational networks, emphasizing the increasing power of using 3D technology for heritage preservation in ethnographic collections.

It has addressed the role of digitization in how institutions might deal with collections from a colonial context. The conclusion underlines the implications of 3D artifact creation, which go beyond mere display and influence curators' strategies for collecting and conserving artifacts. It highlights how the use of 3D modeling should improve efficiency rather than replace human creativity.

The research addresses urgent issues, emerging methodologies, and interdisciplinary aspects of museums and the use of 3D modeling technology. It promotes the collaboration between museum curators and Indigenous artists in artifact selection, incorporating historical records and digitalized artifacts. It advocates for using 3D technology in ethnographic heritage preservation, emphasizing collaboration and outreach.

Through consultations, museums can form meaningful relationships with source communities and should respect community decisions regarding digitization, access, and use, giving Indigenous communities full agency over how their cultural material is used. It showcases the importance of preserving and celebrating cultural heritage, paying attention to their voices, and building relationships based on trust and engagement. The integration of Indigenous knowledge with 3D technology in the Arctic is presented as a pathway for sustainable habitation and human activity in challenging conditions. This chapter aims to stimulate further discussion and contribute to ongoing debates. We highlight key issues and make practical recommendations for policy development, but do not claim to provide definitive answers for all scenarios that may arise.

Acknowledgments

This research could not have been completed without the support awarded to Elisa Palomino from: the Fulbright UK-US scholarship at the Arctic Studies Center, National Museum of Natural History, Smithsonian Institution, and the AHRC LDoc London Doctoral Design Centre Award. This project has received funding from the European Union's Horizon 2020 research and innovation programme under grant agreement No. 823943, *FishSkin: Developing Fish Skin as a Sustainable Raw Material for the Fashion Industry*. Further funding includes an AHRC Kluge Fellowship at the Library of Congress under Grant No. AH/X002829/1. Postdoctoral fellowship at Ca' Foscari University, Department of Asian and African Mediterranean Studies; Max Planck Institute for the History of Science, Department III (Artifacts, Action, Knowledge) in Berlin, facilitated by the staff Ellen Garske, Matthias Schwerdt, Sabine Bertram, and Ruth Kessentini. Postdoctoral fellowship at Kunsthistorisches Institut in Florenz, Max-Planck-Institute, and *Koç University's* Research Center for Anatolian Civilizations, ANAMED. The authors are very grateful for the support from these institutions. The chapter draws from the invaluable insights of Elder June Pardue (Alaska Native, Alutiiq and Inupiaq).

Bibliography

1. Riegels Melchior M, Mackinney-Valentin ME. Introduction: Arctic fashion. *Crit Stud Fashion Beauty*. 2023;14(2):141–54.
2. Hatt G, Taylor K. Arctic skin clothing in Eurasia and America: An ethnographic study. *Arct Anthropol*. 1969;5(2):3–132.
3. Svensson TG. Clothing in the Arctic: A means of protection, a statement of identity. *Arctic*. 1992;45(1):62–73.
4. Konola S, Kähkönen P. *Arctic Wears – Perspectives on Arctic Clothing*. Rovaniemi: Lapin AMK; 2015.
5. Usenyuk-Kravchuk S, Akimenko D, Garin N, Miettinen S. Arctic design: Basic concepts and practice of implementation. In: Jokela T, Coutts G, editors. *Relate North: Tradition and Innovation in Art and Design Education*. Viseu, Portugal: InSEA Publications; 2020.
6. Zhozhikov AV. Digitalization of the cultural heritage of the indigenous peoples of the Arctic. In: Savchenko I, editor. *Freedom and Responsibility in Pivotal Times*. Vol. 125. Eur Proc Soc Behav Sci. Nizhny Novgorod, Russia: European Publisher; 2022. pp. 947–56.
7. Palomino E. Indigenous Arctic Fish Skin Heritage: Sustainability, Craft and Material Innovation [dissertation]. London: University of the Arts; 2022 [cited 2023 May 30]. Available from: https://ualresearchonline.arts.ac.uk/id/eprint/20124/
8. Klobe T, Thomas M. Biocultural collections: Exhibition concept, planning and design. In: Salick J, Konchar K, Nesbitt M, editors. *Curating Biocultural Collections: A Handbook*. Kew: Royal Botanic Gardens; 2014. pp. 373–91.
9. Csoba DeHass M, Hollinger E. 3D heritage preservation and indigenous communities in the circumpolar north. In: Heininen L, Exner-Pirot H, editors. *Arctic Yearbook 2018: Arctic Development in Theory & In Practice*. Akureyri, Iceland: Northern Research Forum; 2018. pp. 218–33.
10. Haakanson S. Translating knowledge: Uniting Alutiiq people with heritage information. In: Silverman R, editor. *Museum as Process: Translating Local and Global Knowledges*. London: Routledge; 2015. pp. 123–9.
11. Wallace A. Decolonization and Indigenization [Internet]. OpenGLAM; 2021 [cited 2024 Jan 13]. Available from: https://openglam.pubpub.org/pub/decolonization/release/1
12. Fraser J. Beyond Copyright: The Ethics of Open Sharing [Internet]. Creative Commons – We Like to Share Medium Blog; 2021 [cited 2024 Jan 13]. Available from: https://medium.com/creative-commons-we-like-to-share/beyond-copyright-the-ethics-of-open-sharing-a495bb95569d
13. Christen K. Access and accountability: The ecology of information sharing in the digital age. *Anthropol News*. 2009 Apr;50(4):4–5.
14. Krupnik I. Recaptured heritage: Historical knowledge of Beringian Yupik communities. *Arct Stud Cent Newsl*. 2000; 8:9–10.
15. Wallace A. Open and Equitable – Questions on How to Handle Digital Copies in Collections from Colonial Contexts (Part III) [Internet]. Wikimedia Deutschland Blog; 2022 [cited 2024 Jan 13]. Available from: https://blog.wikimedia.de/2022/05/18/open-and-equitable-part-iii/
16. Crowell A, Worl R, Ongtooguk PC, Biddison D. *Living Our Cultures, Sharing Our Heritage: The First Peoples of Alaska*. Washington, DC: Smithsonian Books; 2010.
17. Palomino E, Pardue J. Alutiiq fish skin traditions: connecting communities in the COVID-19 era. *Herit*. 2021;4: 4249.
18. Hickman P. *Innerskins and Outerskins: Gut and Fish Skin*. San Francisco: San Francisco Craft Museum; 1987.
19. Meade M. Sewing to maintain the past, present and future. *Études/Inuit/Studies*. 1990;14(1–2):229–39.

20. Fienup-Riordan A. *The Way We Genuinely Live: Yuungnaqpiallerput—Masterworks of Yup'ik Science and Survival*. Seattle: University of Washington Press; 2007.
21. Hurcombe LM. *Perishable Material Culture in Prehistory: Investigating the Missing Majority*. London: Routledge; 2014.
22. Forbes RJ. *Studies in Ancient Technology*. Leiden: E.J. Brill; 1955.
23. McKennan RA. The Upper Tanana Indians. *Yale Univ Publ Anthropol*. 1959; 55:1–223.
24. Snow JH. Ingalik. In: Helm J, Sturtevant WC, editors. *Handbook of North American Indians*. Vol. 6, Subarctic. Washington, DC: Smithsonian Institution; 1981. pp. 602–17.
25. Hadleigh West F. The Netsi Kutchin: An Essay on Human Ecology [dissertation]. Baton Rouge: Louisiana State University; 1963.
26. Osgood C. Contributions to the ethnography of the Kutchin. *Yale Univ Publ Anthropol*. 1936; 14:3–198.
27. Slobodin R. Kutchin. In: Helm J, Sturtevant WC, editors. *Handbook of North American Indians*. Vol. 6, Subarctic. Washington, DC: Smithsonian Institution; 1981. pp. 514–23.
28. Palomino E, Rahme L. Indigenous Arctic fish skin – A study of different traditional skin processing technology. *J Soc Leather Technol Chem*. 2021;105(2): 59–71.
29. Chugachmiut Cultural Heritage & Resources. Learn the waterproof stitch with gut [Internet]. 2023 [cited 2024 Jan 10]. Available from: https://chugachheritageak.org/pdf/CLO_6-12_LEARN_THE_WATERPROOF_STITCH_WITH_GUT_Final.pdf
30. Birket-Smith K. *The Caribou Eskimos: Material and Social Life and Their Cultural Position*. Vol. 1. Copenhagen: Gyldendalske Boghandel, Nordisk Forlag; 1929.
31. Oakes JE, Riewe R. *Alaska Eskimo Footwear*. Fairbanks, Alaska: University of Alaska Press; 2007.
32. Issenman BK. *Sinews of Survival: The Living Legacy of Inuit Clothing*. Vancouver: UBC Press; 1997.
33. Ewing ER, Darwent CM. Sinew thread production and stitch properties in arctic Alaska clothing construction. *Alsk J Anthropol*. 2018;16(2):1–23.
34. Lincoln T, et al. *Arctic: Culture and Climate*. London: Thames and Hudson; 2020.
35. Hollinger RE. 3D digital replication: Emerging cultural domain for Native American communities. In: Krupnik I, editor. *Handbook of North American Indians*. Vol. 1. Washington, DC: Smithsonian Institution; 2023.
36. Clo3D [Internet]. 2021 [cited 2024 Jan 4]. Available from: https://www.clo3d.com
37. Blender [Internet]. 2021 [cited 2024 Jan 4]. Available from: https://www.blender.org/
38. Huang S, Huang L. CLO3D-based 3D virtual fitting technology of down jacket and simulation research on dynamic effect of cloth. *Wirel Commun Mob Comput*. 2022;2022:5835026. p. 1–11.
39. Moskvin A, Wijnhoven MA, Moskvina M. The equipment of a Germanic warrior from the 2nd–4th century AD: Digital reconstructions as a research tool for the behaviour of archaeological costumes. *J Cult Herit*. 2021; 49:48–58.
40. Palomino E, Karmon A, Topaz O, Solo A, Cordoba A. Digital tools in the fashion industry: Fish skin garments and Ainu fish skin traditions. In: Schramme A, Verboven N, editors. *Technology, Sustainability and the Fashion Industry: Can Fashion Save the World?* 1st ed. London: Routledge; 2024.
41. Brown D, Nicholas G. Protecting indigenous cultural property in the age of digital democracy: Institutional and communal responses to Canadian First Nations and Māori heritage concerns. *J Mater Cult*. 2012;17(3):307–24.
42. Quintavalla A, Temperman J, editors. *Artificial Intelligence and Human Rights*. Oxford: Oxford University Press; 2023.
43. UNESCO. Recommendation on the ethics of artificial intelligence. SHS/BIO/PI/2021/1 [Internet]. 2021 [cited 2023 Dec 20]. Available from: https://unesdoc.unesco.org/ark:/48223/pf0000381137_spa

44. United Nations. The United Nations Declaration on the Rights of Indigenous Peoples (UNDRIP) [Internet]. 2017 [cited 2024 Jan 13]. Available from: https://social.desa.un.org/issues/indigenous-peoples/united-nations-declaration-on-the-rights-of-indigenous-peoples
45. Gonzalez Zepeda LE, Martinez Pinto CE. Regulation of Advanced Digital Technologies for Indigenous Material Culture in the Museum Heritage Landscape [Internet]. UNESCO Office Montevideo and Regional Bureau for Science in Latin America and the Caribbean; 2023 [cited 2024 Jan 4]. Available from: https://unesdoc.unesco.org/ark:/48223/pf0000387814
46. Settee P. Indigenous knowledge: Multiple approaches. *Counterpoints*. 2011; 379:434–50.
47. Lewis JE, et al. Indigenous protocol and artificial intelligence position paper. Project Report. Indigenous Protocol and Artificial Intelligence Working Group and the Canadian Institute for Advanced Research; 2020.
48. United Nations. 2030 Agenda for Sustainable Development [Internet]. 2015 [cited 2024 Jan 4]. Available from: https://www.itu.int/en/ITU-D/Statistics/Pages/intlcoop/sdgs/default.aspx
49. Benedict C. Traditional Knowledge and Copyright Intersections [Internet]. Creative Commons – We Like to Share Medium Blog; 2023 [cited 2024 Jan 13]. Available from: https://medium.com/creative-commons-we-like-to-share/traditional-knowledge-and-copyright-intersections-d8cb78375e40
50. Vézina B, Muscat A. Sharing Indigenous Cultural Heritage Online: An Overview of GLAM Policies [Internet]. 2020 [cited 2024 Jan 13]. Available from: https://creativecommons.org/2020/08/08/sharing-indigenous-cultural-heritage-online-an-overview-of-glam-policies/

13 At the River—Stepping in the Flow

Timo Jokela

13.1 Introduction

I borrow the well-known notion "You cannot step into the same river twice" from the Greek philosopher Heraclitus for the title of the chapter. In a way, rivers are eternal flows, but they are also in a state of change; the seasons, amount of water, light, wind, and other factors affect the river over time. Heraclitus clarifies his observation by saying, "If you step into a river, the flow is no longer the same, nor are you as you were before." This chapter covers a time window of approximately 50 years. The river I examine through my art has changed, but so have my thoughts, perspectives, paradigms, and the concepts I use, reflecting the meaning of the river. The title also refers to one of my latest artworks related to the river, which I will return to at the end of the chapter.

The river in focus in this chapter is the Ounas River, a tributary of Kemijoki, the largest river system in Finland. Köngäs is my home village, where I grew up and where I still live part-time. Originally, the river was an important route, the "main street" of the village, along whose banks the houses were built in a deliberate and practical connection with the river. The river was the core of the village's way of life. The river's ecosystem and people's lives, livelihoods, and social and cultural practices are intertwined in a way that can be called ecoculture [1, 2]. The concept of a "taskscape," as defined by Ingold [3], aptly describes the life and work of the villager in relation to the river landscape and its temporary changes. The concept of a "taskscape," as defined by Ingold, aptly describes the life and work of the villager in relation to the river landscape and its temporary changes. The river landscape extends beyond visual interpretation, emphasizing temporal dimensions and lived experience. Various tasks, such as haymaking along riverbanks, fishing according to the seasons, log floating, and transportation along ice roads, connect people to natural cycles and create a dynamic flow of lived activity. Taskscape is Ingold's way of explaining how people actively create, inhabit, and experience their environments, not just as static spaces, but as dynamic fields of practice and interaction.

Throughout my relatively long career, I have witnessed the changes people living by the Ounas River have faced. A river plays a significant role in the construction of a village's identity. The chapter is a review, where the research data

DOI: 10.4324/9781003545880-17

serve as the objects of Baradian diffractive reading. This is an approach that acknowledges the interconnectedness of different perspectives, discourses, and entities. In essence, Baradian [4, 5] diffractive reading challenges the idea of fixed, stable meanings and engages with texts and ideas in a way that acknowledges the complexity, multiplicity, and ongoing dynamics of knowledge production [4, 5].

In my chapter, I highlight three phases of large changes and threats to the river and the lives of the river's community. Through my art related to rivers, these phases could be seen as new beginnings and perhaps even as a new paradigm in artistic work related to rivers. These phases include the threat of hydropower construction and reservoir dams, the end of two significant livelihoods—milk production and log floating—and, finally, the threat of the global mining industry to the water system.

The chapter illuminates entanglements and the negotiation of tensions between Western art and Northern knowledge [2, 6], humanistic and social place-specific approaches [7], and posthumanistic tension knowing-with other species [8, 9]. Finally, I link my chapter to the current discussions on the new genre Arctic art [2, 10] and its possible role as a leverage point in Arctic sustainable transformation [11, 12].

13.2 Threat of Hydropower Dams and Water Reservoir Lakes

Since my childhood, the river has been a constantly changing stage for natural phenomena, a stage where the village lives its daily life, culture, and social commitments. In addition to being a natural ecosystem, a river is a powerful and influential aesthetic factor that affects the culture of a village in many ways. In my dialect, there are hundreds of words for the manifestations of the changes of river and water throughout the year; most of these concepts are amalgamations of Finnish and the now-extinct Kemi-Lapland Sámi language spoken in the region before. These concepts are proofs and expressions of villagers' living ecoculture, in which the river still plays a central role.

Memories of my childhood's happiest moments include those spent by the river in connection to the flow of water and engaged in various ecocultural tasks, such as fishing, haymaking at the riverbanks, learning to paddle, or just following the flow as it reflected the lights of the sky. However, even in my childhood in the late 1960s, the Ounas River and my home village faced threats. In my elementary school geography textbook, there was a depiction of Lapland county, a map with a massive artificial lake planned for hydroelectric power generation situated to the right of my home village. The text next to the map describes the typical colonization of the North at that time. It read, "Living conditions are difficult in Lapland, much help from the south is still needed there." This was the learning lesson I received in school. Later, I understood that we, the children of the North, were educated to leave our homes because, in our villages, we were an obstacle to resource-exploiting modernization.

The village lived in an uncertain situation, being afraid for decades. On the neighboring rivers, the large water reservoirs of the other tributaries of the

Kemi River had just been built, and the villages were flooded. People were relocated to new places, almost using aggressive force. The southern region needed more hydroelectric power for industry. Kaulanen [13] demonstrated how the construction of hydropower supported by the state aimed to spread the view that the areas submerged underwater were empty, uninhabited wastelands. On the other hand, conservationists from Southern cities wanted to defend the almost untouched, valuable nature of the rivers [13, 14]. For the locals, it was a matter of their fate: their homes, livelihoods, culture, and future.

As a teenager, I had the fortune to encounter some of the first professional artists living in Lapland. As a boy, I dreamed of being a hunter and fisherman; instead, I was given an alternative: to make art. I saw that art was real work because the local artist was a respected man in his community. The school taught things and skills needed in cities and aimed to educate me to leave behind the remote village and the ties it had to the land and water, which were seen as the opposite of modernization and civilization. Art creates the possibility of preserving rivers in a new way. I began painting with the intention of capturing the landscape and life along the river. This led me to pursue art studies in Helsinki. I am grateful to the artists and to my river for that opportunity.

At the same time, there were strong protests, Ellos Deatnu! movement in the Sámi region in Alta, Norway, against the construction of a hydroelectric plant. The Sámi people defended their rights to the land [15, 16]. This made me increasingly realize the significance of the river and its essence as a part of my culture, which was a part of me. I understood then that protecting rivers was also protecting our culture. We shared stickers that read, "Water is the Alpha and Omega of life. Did you know that the Alta and Ounas Rivers have common spring water?" I realized that it was not just about the water system but the sources of our culture, too. What I began to outline back then as the core of my art related to the river can now be examined through the concept of ecoculture, a nexus where the river and humans shape each other and manifest as an Ingoldian taskscape.

Despite the massive Ellos Deatnu! protest, the Alta hydroelectric power plant in Norway was built with the support of the largest police operation in Norway since World War II. However, the struggle reinforced Sámi identity and directed new cultural strength to the North. The Alta protests marked the revitalization of Sámi yoik music and its transformation into concert-style events powered by young performers. The foundation of Sámi art had been traditional handicrafts, duodji, which now received alongside it a new form of expression based on contemporary visual art and the significant Mázen art group (sámi dáiddárjoavku) was founded in close association with the Ellos Daetnu movement in 1978. For art, it was even necessary to create a completely new concept "dáidda" in the Sámi language [15, 16]. It was meaningful for me to follow what was happening in Norway among the Sámi people. In those days, in the 1970s, I started step by step to understand what colonization of the North and its people's minds meant for the in-home regions. For me, the question was not so much about ethnic colonialism, but more about the colonialism of the rural North.

13.2.1 Painted Ecocultural Taskscapes at the River

Shortly after Alta, in the late 1970s and 1980s, the Ounas River was in danger. Dams and water reservoirs were planned to be built on the Ounas River [13, 14]. I was involved in activities against this. I was also looking for ways to incorporate my own art into this battle. To show that the land threatened by hydropower construction was not an "empty wilderness" but inhabited by a valuable living culture, I started making landscape paintings and holding exhibitions not about the wilderness river but instead focused on the ways of life and vernacular architecture along the riverbanks.

In hindsight, this does not seem very innovative, but for me, it was a conscious choice. Having just started studying art in Helsinki, I had been guided toward abstract art, which relied on the idea of the universality of modernism. My interest in the landscape, nature, and the North was seen at the Southern art school as a sign of dilettantism. On the other hand, Lapland and the North in general had, until then, mostly been depicted through the eyes of visiting artists, who wanted to see Lapland as an untouched wilderness, without the people living there and their traces in the landscape.

Now, nearly 50 years later, when reviewing my paintings and exhibitions, it becomes clear that even traditional cultural landscape paintings have significance. By presenting villagers' riverside taskscapes, they helped in recognizing the value of Northern ecocultural heritage, especially during discussions and negotiations about it (Figure 13.1). This, in turn, strengthened the sense of community, which can be seen as a great resource for defending the river. The Ounasjoki Movement won the battle, and the hydroelectric company and the Finnish state gave up. In 1983, Ounasjoki was protected by law [13, 14]. The river was secured, or so we thought.

13.3 Toward Knowing-with the River

I see the struggle to protect the Ounas as the main reason that I became interested in cultural history, anthropology, humanistic geography, and interdisciplinarity

Figure 13.1 Timo Jokela's oil paintings from the late 1970s depict the strong ties of the village of Köngäs to the Ounas River. Photos: Timo Jokela, 2023.

[17, 18]. I began to study the river using a place-based research approach. I studied art in the South in the capital city of Helsinki and taught art and art education at the University of Oulu, but I did not abandon my river in the North.

In contrast, after being freed from the restrictive and colonizing influence of modernism, art provided new meaning to my experiences of living with the Northern river and new tools for exploring the river. The exhibition titled *MA, space and time* in Japanese art and architecture helped me realize the sensitive and precise observations of nature that I had learned in the ecoculture of my home village. Japanese art's meditative approach to nature mirrored my Northern knowledge and experiences in fishing, hunting, and berry picking, helping me understand dimensions of perception, now referred to as embodied knowledge, embodied learning, or land-based learning. This offered a new perspective on my interest in the Ingoldian taskscape and ecocultural entanglements with the river [17, 18].

The Arte Povera exhibition I saw in Paris reconnected me with traditional materials and working methods and inspired me to use natural materials in my work. This led me to become an environmental artist who often used natural materials from the river landscape for installation. Today, there is growing interest in material culture and handmaking skills in art and art education, reflecting a global shift toward new materialism and emphasizing the importance of materiality in the art of the Arctic and Northern regions [7, 19].

As I moved from landscape painting to place-specific art, the meanings given to the river by those living along it and the stories associated with it became emphasized in my artistic work. The visual aspect of my work was complemented by strong humanistic and sociocultural dimensions. I challenged myself to develop artistic activities that highlight the river's ecoculture, Northern knowledge, and the ways of knowing-with the river, that is, with its flow. Today, the concepts are linked to posthumanism. At that time, I theorized about the change through the concepts of community-based environmental art and environmental performances, as well as Lacy's [20] new genre of public art, which she defined as socially engaged practice happening directly with participants outside the art world, and Lippard's place-specific art [21] that occurs also in small rural communities and their places outside of city centers. This chapter reveals how, about 30 years ago, I outlined the starting point of my art as what my colleagues and I now theorize and codevelop under the concept of new genre Arctic art [2, 22].

In 1994, I moved back to the North to establish a new Faculty of Arts at the University of Lapland in Rovaniemi, a town located at the Polar Circle. With my new colleagues, I immediately initiated a multidisciplinary project called "River" (*Joki* in Finnish). It was a combination of the environmental artworks on sites, exhibitions at the Rovaniemi Art Museum, and the Summer School at the University [17, 23]. My goal was to demonstrate and test my arts-based, multidisciplinary thinking. For me, the obvious target was the Ounas River. In the lecture series at the University, the river was approached from various disciplinary perspectives, and we also organized field trips to villages to meet and listen to the people.

Reviewing the results 30 years later, I see how the river appeared to the residents as an Ingoldian taskscape, with its social dimensions intricately linked to the river's ecoculture. Two aspects prominently stood out in discussions with the villagers. The older generation had warm memories of haymaking along the riverside meadows outside the village, together with their families. The other significant aspect, especially for men, was the log floating run by lumber companies and the forest industry. Log floating stopped in the early 1990s because of pressure from the hydroelectric power company as they had managed to buy the river's flow—hydropower—through lobbying.

Log floating played a significant role in my artistic work during the Summer School of 1995. The riverbank meadows and haymaking, however, had their role in my art only later. Together, they form a pair through which I depict changes in both the river and my art, especially in their entanglement and intra-action.

13.4 River Flow—Log Raft as Community Art

For the men in the village, the river was a key transportation route. They specialized in transport, using boats in the summer and river ice as a winter road. Children practiced paddling on the river under the supervision of adults, and young people competed with each other and learned to read the flow of the water. The question of Northern knowledge [2, 6] shifted from generation to generation. The river was used for log floating for 100 years. At the age of 15, boys joined log-driving teams composed of village men, which can be seen as a rite of passage into adulthood. Work on the river was secured by fathers, uncles, or other familiar men from the village—those who had knowledge of the flow of the river. Interestingly, my investigation in the mid-1990s showed the deeper meaning of log floating for villagers. When it ended, people mourned not the loss of income but the disappearance of an important social, communal, and even aesthetic event the river had provided [17].

When I started the River Summer School, I designed a workshop for students and staff where I aimed to demonstrate how the river should be approached and studied according to my understanding. With my colleagues, we examined the rafts used to transport logs to the river mouth in hundreds of log bundles and borrowed 30 logs from the forestry agency to construct the raft (Figure 13.2). We launched it down the river, more than 200 kilometers, met the villagers, and documented their stories and lives with the river. We were hailed as kind of heroes everywhere we went because people said that we brought the log floating back to the river. We created environmental art installations along the river using gifts and materials. The project concluded with an exhibition at the Rovaniemi Art Museum, a part of which was also exhibited in Montreal, Canada.

For me, the log raft performance was significant in many ways. Through it, I brought myself back to the river as an artist to intimately learn from its flow and essence. I felt I had succeeded in incorporating Northern ecocultural knowledge and ways of knowing into my art. Today, nearly 30 years later, with new and refined concepts, I can better argue for the paradigmatic impact of log

Figure 13.2 (Top four images) The first moments of the log raft performance. Photos: Timo Jokela, 1995. (Bottom two images) Snow installation in front of an old haybarn and installation in art gallery telling stories of haymaking taskscapes at the Ounas River. Photos: Timo Jokela, 2022 (bottom left) and Marko Junttila, 2023 (bottom right).

raft performance and its ontological and epistemological significance not only for art but also for encounters between Northern ways of knowing and the Western academic world.

Another significance relates to art education and teaching art at the University. Through the log raft event, I felt I had successfully demonstrated a new paradigm of art education to my colleagues and students, not only in art education but also in other disciplines at the University. Despite this, it has taken time to integrate ways of thinking into the structures of academic subjects and curricula. This is the challenge we work under within the initiative called New Genre Arctic Art Education. In one way, I feel that I am still on the same development flow as log rafts on Ounas 30 years ago.

13.5 Vanishing Meadows—Art of the Haybarns

During the COVID-19 pandemic, I was given the opportunity to return to my river and realize a series of artworks that I had long dreamed of. "Living in the landscape" was a course I had developed with my colleagues in international collaboration at the University of the Arctic. The course was held annually in various parts of the Arctic region, focusing on teaching art-based landscape research methods. At the time of the pandemic, we did not gather together; instead, in addition to the online course, everyone worked in their own home landscapes. My landscape was river meadows in April, which allowed me to rethink the village's taskscapes at the river. I worked on the river, searching for fragments of the lost meadow economy and witnessing changes in the riverside nature [24].

As ecocultural taskscapes [1, 3], riverside meadows have a long history, and haymaking has affected river landscapes and ecosystems for centuries. The village was originally established because of the fertile hay meadows along the river, which are essential for feeding domestic animals such as cows, horses, and sheep. Every household could have meadows far from the village on the riverbank, its islands, small tributaries, and creeks. The hay meadows were originally natural floodplain meadows or old beaver dam ponds. Later, they were cleared along the riverbanks into kilometer-long, fully flowering meadows. The small wooden haybarns, of which there were hundreds along the river, served as shelters for the hay and the haymakers working on the site. Common haymaking was a time of gathering, togetherness, and celebration of life in summer—a central taskscape for all village families. Haymaking was working along the river in which everyone, except mothers with infants and the immobile elderly, participated. The long history of haymaking shaped villagers' attitudes toward the river and even their identities. By the 1980s, riverside meadows were abandoned because horses were replaced by tractors that could not access meadows across the river. Today, nature has almost reclaimed its own: The open meadows have become overgrown, and the plants, insects, and birds of the meadow ecosystem have almost disappeared.

In the April–spring snowdrifts, when the river was still frozen, I wandered among the old meadows far away from villages and searched for barns still standing, the oldest dating back to the 1700s. The barns were gathering places, especially for young people; some even spent nights there. They were their own versions of modern-day social platforms. Their walls were filled with writings. The oldest, carved with a knife 200 years ago, the latest, written with a pencil last year. There are many entries about weather, rain, river water levels, fish, birds, relationships, infatuations, falling in love, the world's pain, and miserable fates.

Snow and ice art have been my materials, and their changes according to the weather have been an important method for working at the river in winter. I collected photographs of people who had worked in the barns, installed the photos inside ice blocks, spent moments with them, and read their carvings and writings on the wall of the haybarns (Figure 13.2). By documenting snow and ice sculptures, repairing found, old tools, and copies of the wall carvings of riverside barns, I have created installations for various exhibitions. Installation of artworks and connected everyday materials have opened up new possibilities for examining the relationships between local craftsmanship, design, and art, which Western art has separated from each other since the Enlightenment. In contrast, Arctic art has been seen as intertwined [11, 22].

13.6 Stepping in the River

The next artwork through which I explore the relationship between my art and the river is an installation, including a video, that shares the same name as this chapter. The artwork was displayed at the Korundi Rovaniemi Art Museum as part of the Shifting Ground exhibition in the first part of 2024 [25]. The exhibition was the result of an art and research project in which I was involved. The Shifting Ground project was a Canadian-Finnish collaboration coordinated by Emily Carr University, Vancouver, Canada.

Into the artwork, I integrated observations from my home river, the Ounas and Yukon Rivers, and its tributary, the Klondike River, in Canada. The key commonalities between these two rivers are the similarities in nature and ecosystems and the resulting similarities in the ecocultures of people in Yukon and Lapland in Finland. Another common aspect is gold mining and the impact of the mining industry on water systems and people living with these lands and waters. The gold rush in Yukon, Canada, began over a hundred years ago. According to Dobrowolsky [26], the Han people had to make way as a town of gold miners rose in place of their fishing village at the mouth of the Klondike River, a tributary of the Yukon River. The Canadian-owned Kittilä Gold Mine was established in 2006 near my home village in Lapland, within the Ounas River water system.

In Yukon, Chief Isaac, the leader of the Han people, tried to protect his people and their way of life in the event of inevitable change. According to Dobrowolsky [26], Chief Isaac handed over the ritual items of his people to a neighboring tribe, hoping to get them back if they survived the situation. To protect their community, he also moved the village to a new location called Moosehide, which is currently a place for famous cultural gatherings of the Yukon River Han People. During my visit to Yukon, I understood the similarities their culture and that of my home village have had because of the similarities in nature—deer hunting, fishing, berries—the same activities following the cycles of nature as I have met and learned in Lapland, Finland.

During my visit to Yukon, I had a profoundly emotional and enlightening moment when I saw an old photograph. Taken approximately 1900, the photograph depicts Chief Isaac ice fishing for grayling on the Yukon River in front of his Moosehide village. It struck me deeply because he was sitting on river ice in the same familiar position as my father in a photo I took in the early 1970s, where he was also ice fishing for grayling at our river.

The grayling, a site-faithful fish that remains almost in one place throughout its life, was the most valued fish of my generation. By the late 1940s, migratory fish, such as salmon, had already disappeared from our river because of the construction of hydropower dams at the mouth of the Kemi River. Grayling was almost a daily food that we had learned to fish from the river since childhood. The river's water level and temperature regulate the lives of grayling, so the river is followed by "knowing-with" grayling, wondering if grayling had risen to the rapids, whether it eats the fly now, and so forth. The beauty of the river's flow was admired by saying, "This is a good place for grayling to live." Especially for men, it is common to express aesthetic experiences through "knowing-with" fish and animals [17].

I saw Chief Isaac's photo in 2019; around the same time, I learned about the construction of a 16-kilometer-long wastewater discharge pipe by the Kittilä Gold Mine. This pipe was directed to the Loukinen River, a main tributary of the Ounas River at the mouth of the Sotkajoki River, which is known as a spawning and ice fishing area for grayling in the village. This construction occurred even though the Ounas River water system was legally protected. The mining company had not yet received permission for construction, but later, the mine was granted approval. Those who approved the wastewater discharge pipe construction argued that the law does not mention mines, hydropower, and other industrial constructions only. According to the documents, the law's intent was to protect the river from all changes, but mines were forgotten to mention. Ethically and morally, this oversight is profoundly wrong, illustrating how Northern colonization continues legally. In the North, especially among the Sámi, the green transition is viewed as a new form of global "green colonialism" [27]. Just as during the earlier hydropower construction, the North is perceived as an area devoid of inhabitants and ripe for exploitation in the name of global economic growth.

These reflections inspired my artwork in the Shifting Grounds exhibition. Although my work is grounded in place studies within its humanistic, historical, and anthropological context in the spirit of visual anthropology, it also has an activist, decolonizing undertone. I share Arora-Jonsson's [28] notion of the importance of moving from interdisciplinary integration toward embracing ambivalence and humility, opening up to experimental practices and "trading places" to engage justly with nature and others.

Through my artwork, I tell a story about rivers and people's ecoculture in situations of change and threat. The central idea is knowing-with grayling, which requires free-flowing, clean, and sufficiently cold water to survive. In this video of the indoor installation, I have woven together documentation of my work in Yukon with abandoned gold mining hoses and the construction of the Kittilä Gold Mine wastewater pipe, photos of Chief Isaac and my father and me of ice fishing for grayling. As the core of the video in ice installations, I created the site where the wastewater from the gold mine flows into the Loukinen River. Grayling swimming inside ice blocks are shown as transparent photos (Figure 13.3).

The Loukinen River itself, winding and meandering through the swampy area, inspired the shape and rhythm of the material part of the installation exhibited in the art museum. As the son of a boat maker, the curves and bends of trees have always been an endless source of inspiration for me. For this

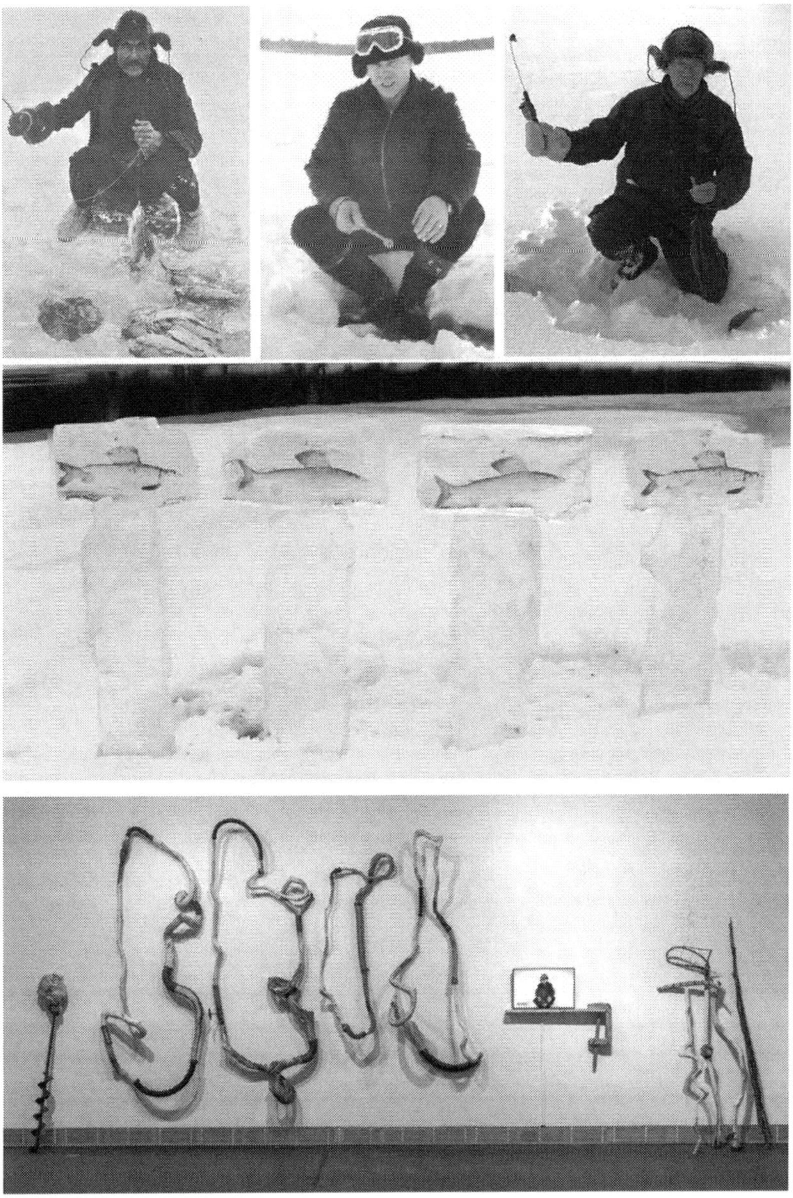

Figure 13.3 I combined in my video installation photos of Chief Isaac, my father, Aatu Jokela, and my own grayling ice fishing photos. I edited them together with the construction of the mine's discharge pipe, my artistic work in the Yukon goldfields and on ice installations with graylings at my home river. Photos: Wikimedia Commons/University of Washington Digital Collections (above left), Timo Jokela, early 1970s (above middle), Mirja Hiltunen, 2022 above (right). Details of the ice-installation at the site where wastewater ends up in the Loukinen River, Timo Jokela, 2023 (middle). Timo Jokela's installation with video *At the River—Stepping in the Flow* at Korundi Art Museum. Photo: Tatu Kantomaa, 2024 (below).

work, I collected wood material from the riverbanks with my grandson. We found pieces of rotten pipes and old tools, along with people who have been connected to the river's nature. These items in the installation were not brought from the gold mining fields in Yukon, though one might think so when looking at the video accompanying my art piece.

13.7 *The Ounas – Our Free River*

The last case through which I explore the changes at the river is a public artwork created collaboratively with the villagers of Köngäs [29]. This work has its origins in the late 1970s and early 1980s, during the movement to protect the Ounas River from hydroelectric construction. To memorialise the successful activism described earlier, the villagers wanted to erect an artwork as a tribute to the river.

The visual elements of the artwork, which I designed together with the villagers, reference the residents' coexistence with the river. The villagers participated in various ways in creating the artwork: by sharing stories, applying their skills during the construction, and celebrating the final result. Old photographs, behind the doors, inside the artwork's "memory box," and the stories shared around the artwork reflect many meaningful aspects of the local ecoculture, traditions, and changes of living with the river. Implementing the project together with the villagers and the river, the public artwork can be seen as a self-portrait of the community and as the community's ability to respond to change (Figure 13.4).

The foundation of the artwork is the villagers' relationship and intra-action with the river as a key element of their ecoculture—the river as a taskscape—through which the village's identity has been shaped. Art that integrates visual arts, crafts, and landscape architecture—while connecting Northern ecocultures and identities—plays a vital role in discussions about revitalization, resilience, and sustainability in rural Northern communities. As "a sustainability portrait" of the village, the artwork is a statement in support of preserving the river's cleanliness in the face of threats from mining activities.

13.8 Conclusion

In this chapter, I have discussed my artwork entanglements with respect to the ecoculture and Northern knowledge of riverside villages. The chapter is also a reflection on my own roots as an artist and a northern man. The latest visits to Yukon in Canada acted as a mirror, deepening my understanding of the meaning of rivers for Northern cultures. It is possible the visit also provided a deeper understanding of Heraclitus's multilayered message: "You cannot step into the same river twice; if you try, the flow is no longer the same, nor are you as you were before." The river has changed and has been a great teacher for me since my childhood. Currently, toward the end of my academic career, I have been exploring the river through both posthumanism's knowing-with and Indigenous pedagogy's land-based learning approaches.

Figure 13.4 Timo Jokela's artwork, *The Ounas – Our Free River*, designed and created with the villagers, serves as a portrait of the river and the village. The visual elements of the piece—old photographs, gears related to life with the river, and texts within the "memory box"—along with the stories told during the process and celebrations, reflect many meaningful aspects of the local eco-culture and traditions. Photos, Timo Jokela, 2024.

Today, the most relevant perspective for me is the concept of new genre Arctic art, something my colleagues and I have defined. I see this concept as having emerged "in the flow of the river" gradually over time. With this concept, we examine the relationship between art and the Northern and Arctic regions, as well as how art can contribute to the transformation of Arctic sustainability. We created this concept to define and describe contemporary artistic interventions, participatory public art, and performances that include activism and engagement with contemporary issues, such as the extraction of natural resources and the cultural rights of indigenous and nonindigenous people.

The new genre Arctic art can be seen as a pedagogical turn in contemporary art as artists have begun to share new understandings of the North and the Arctic with wider audiences. Working with the river, I have always understood my activities as art education and have explored how art can exceed educational frameworks, foster critical dialogue, inspire action, and deepen connections to the rivers. I understand the potential of art to leverage Arctic sustainability transformation precisely in its ability to bring forth and provide alternate means of learning and ways of knowing.

As I write this chapter, the Ounas River still flows, but the generations whom I interviewed about their "northern ecocultural knowing-with the river" at the turn of the millennium are now almost all gone. That is why I am especially happy that I was able to carry out a latest project with the current villagers to create a community-based public artwork called *Ounas – Our Rivers* along the riverbank of my home village.

References

1. Pretty J. Interdisciplinary progress in approaches to address social–ecological and ecocultural systems. *Environmental Conservation*. 2011;38(2):127–139. doi: 10.1017/S0376892910000937
2. Jokela T, Hiltunen M. New genre arctic art education as a way of knowing with the North. In: Jokela T, Huhmarniemi M, Burnett K, editors. *Relate North*. InSEA Publication; 2024.
3. Ingold T. The temporality of the landscape. *World Archaeology*. 1993;25(2):152–174.
4. Barad K. Diffracting diffraction: cutting together-apart. *Parallax*. 2014;20(3):168–187. doi: 10.1080/13534645.2014.927623
5. Barad K. Nature's queer performativity. *Qui Parle*. 2011;19(2):121–158. doi: 10.5250/quiparle.19.2.0121
6. Huhmarniemi M, Jokela T. Arctic arts with pride: discourses on Arctic arts, culture and sustainability. *Sustainability*. 2020;12(2):604. doi: 10.3390/su12020604
7. Huhmarniemi M, Jokela T. Arctic art and material culture: northern knowledge and cultural resilience in the northernmost Europe. In: Heininen L, Exner-Pirot H, Barnes J, editors. *Arctic yearbook 2020: climate change and the Arctic: global origins, regional responsibilities?* Arctic Portal; 2020. https://arcticyearbook.com/
8. Huhmarniemi M, Jokela T, Hiltunen M. Paradigm shifts in northern art, community and environment studies for art teacher education. *Social Sciences & Humanities Open*. 2021;4(1):100181. doi: 10.1016/j.ssaho.2021.100181
9. Ulmer J. Posthumanism as research methodology: inquiry in the Anthropocene. *International Journal of Qualitative Studies in Education*. 2017;30(9):832–848. doi: 10.1080/09518398.2017.1336806
10. Demos TJ. *Decolonizing nature: contemporary art and the politics of ecology*. Sternberg Press; 2016.
11. Jokela T, Huhmarniemi M, Beer R, Soloviova A. Mapping new genre Arctic art. In: Heininen L, Exner-Pirot H, Barnes J, editors. *Arctic yearbook 2021: defining and mapping sovereignties, policies and perceptions*. Arctic Portal; 2021.

12. Priebe J, Lempinen H, Vikström H. *Arctic sustainability transformation 2023*. Umeå University; 2023.
13. Kaulanen J. *Terra Nullius ja paikallisyhteisöt modernisaation puristuksessa 1950- ja 1960-luvun Lapin vesivoimarakentamisessa: Pohjoiskalotin suunnitelmaksi jäänyt rakentamishanke ja Sodankylän säännöstelyaltaat* [*Terra Nullius and local communities in the grip of modernization in the hydroelectric construction of Lapland in the 1950s and 1960s: The construction project that remained a plan for the Northern Lapland and the regulation reservoirs of Sodankylä*]. Lapin yliopisto; 2024.
14. Suopajärvi L. *Vuotos- ja Ounasjoki-kamppailujen kentät ja merkitykset Lapissa* [*The fields and meanings of the Vuotos and Ounasjoki struggles in Lapland*]. Lapin yliopisto; 2001.
15. García-Antón K, Gaski H, Guttorm G, editors. *Let the river flow – an Indigenous uprising and its legacy in art, ecology and politics*. Office for Contemporary Art Norway; 2020.
16. Bjørklund I. The Áltá Action as a Sámi Conflict. Forty years later. In: García-Antón K, Gaski H, Guttorm G, editors. *Let the river flow – an Indigenous uprising and its legacy in art, ecology and politics*. Office for Contemporary Art Norway; 2020. pp. 39–50.
17. Jokela T. Ympäristöstä paikaksi – paikasta taiteeksi [From environment to place – from place to art]. In: Huhtala A, editor. *Ympäristö – arvot? heijastuksia pohjoiseen* [*Environment – values? Reflections on the North*]. Lapin yliopisto; 1996. pp. 161–178.
18. Jokela T. A way to Arctic art education – intersections of land, art and northern knowledge. In: Shin Y, Richards A, Willis S, editors. *The intersectionality of critical identities in art education*. InSEA Publication; 2023.
19. Gamble CN, Hanan JN. What is new materialism. *Journal of the Theoretical Humanities*. 2019;24(6):111–134. doi: 10.1080/0969725X.2019.1684704
20. Lacy S. Debated territory: towards a critical language for public art. In: Lacy S, editor. *Mapping the terrain. New genre public art*. Bay Press; 1995. pp. 171–185.
21. Lippard L. *The lure of the local: senses of place in a multicentered society*. New Press; 1997.
22. Jokela T, Huhmarniemi M, Beer R, Soloviova A. Mapping new genre Arctic art. Arctic Yearbook 2021; 2021.
23. Lohiniva L, Jokela T. *Ympäristötaidetapahtuma Ounasjoen alkulähteiltä Ounasjokisuulle lokakuulta 1994 marraskuulle 1995* [*Environmental art event from springheads to mouth of river Ounasjoki from October 1994 to November 1995*]. Rovaniemen taidemuseo/Lapin yliopisto; 1995.
24. Jokela T. A taskscape behind the barn and at the edge of the forest. In: Härkönen E, Parpala M, Gårdvik M, editors. *Dwellers in the landscape. Lila summer school and exhibition*; University of Lapland, 2022. pp. 42–43.
25. Beer R, Huhmarniemi M. Shifting ground: art exhibition as informal pedagogy on the changing Arctic. In: Jokela T, Huhmarniemi M, Burnett K, editors. *Relate north*. InSEA Publication; 2024.
26. Dobrowolsky H. *Hammerstones. A history of the Tr'ondëk Hwëch'in*. Tr'ondëk Hwëch'in Publications; 2014
27. Hanaček K, Kroger, M. & Martinez-Alier, J. Green and climate colonialities: evidence from Arctic extractivisms. *Journal of Political Ecology*. 2014; 31(1): 538–566. doi:10.2458/jpe.5512
28. Arora-Jonsson S. Does resilience have a culture? Ecocultures and the politics of knowledge production. *Ecological Economics*. 2016;121:98–107. doi:10.1016/j.ecolecon.2015.11.020
29. Jokela T. *Meän vapaa Ounas muisteluteos* [*Our free Ounas – memorial artwork*]. Lapin yliopisto; 2024. https://urn.fi/URN:ISBN:978-952-337-468-3

Conclusion

Antonia Sohns

The challenge of bringing together a diverse array of authors, each engaging with Arctic environmental change through distinct artistic and scholarly perspectives, lies in weaving their varied insights into a cohesive whole. While their approaches differ, ranging from Indigenous ontologies and multimedia storytelling to site-specific interventions and digital reconnections with material culture, the chapters collectively reveal the power of art as both witness and method of response. This collection presents multiple perspectives of Arctic change to illuminate how artistic practices make visible the entangled forces of climate disruption, colonial histories, and cultural resilience. As David Abram described, true knowledge arises in the reciprocity of sensuous exchange between body and land. By that measure, artistic practice is not an add-on to Arctic research—it is a primary mode of inquiry, indispensable for grasping the living dynamics of the changing Arctic [1].

Throughout the book, three main threads emerge. First, art-making is observation. In Junkerdal National Park, Huhmarniemi and Lande entwine art with ecological restoration to explore how artists document, reflect, and comment on changes in nature. Addressing complex questions around restoration processes, Huhmarniemi and Lande described ways of decolonizing Arctic nature and art. Situated noticing and observation also underpin Penney and Allen's participatory photography in Nunatsiavut, where community images of contamination become both dataset and testimony, capturing the emotional texture of environmental change.

Other contributors extended this observational arc. Dunn's paired projects, Raven's Light and Sensing Ice, create a dialogue between sacred, embodied experience, and the abstractions of remote sensing. Gingrich's Siḷamiñ explored weather as breath, asserting Iñupiaq language and cosmology as analytic tools for climate resilience.

These works extend deep listening to lands and waters [2]. They affirm the idea that watching, drawing, and carving, among other arts, are forms of correspondence that move alongside phenomena [3]. Art-making is a rigorous empirical practice: it gathers pixels, roots, antlers, breath, carvings, and stories into composite evidence, demonstrating that to witness Arctic change

DOI: 10.4324/9781003545880-18

responsibly, we must closely observe, listen to, and collaborate with the land, not merely look at it. Art thus expands the evidentiary base of Arctic science, transforming sensing into sense-making.

Second, the book highlights how art mobilizes dialogue to braid multiple knowledge systems and convert observation into collective speech [4]. Hoppu and Isomursu showed how documentary photographs of Northern places can shape viewers' sense of environmental authenticity and underscore how visual culture influences what landscapes mean and how they are valued.

Dialogue includes not only visual perception and communication, but also shared making. Hiltunen and Korsström-Magga worked with Sámi and Arctic youths to carve fears and hopes into reindeer antlers now installed before the Sámi Parliament. Jokela used arts-based action research to create space for community discussion and response to a mine. Solovyeva traced how Sakha artists translate environmental change into paintings that affect local and global audiences. Further, Griebel and Griebel combined anthropological data with allegorical sculpture to open a dialogue about the reach of global consumption in the Arctic. Throughout these chapters, Indigenous and non-Indigenous lenses focus together on environmental change in the Arctic without collapsing difference, using "Two-Eyed Seeing" [5].

Third, this collection emphasizes how art can re-sensitize bodies to land and water by restoring perception through direct sensuous realities of the world [1]. In their chapter, Stöckell and Luostarinen used play in Lapland's forests, story-making, movement, and craft sessions to encourage new sensory and ethical bonds with ecosystems and to create connections with one another. On the Juneau Icefield, Mode mixed glacier-water pigments, data drawings, and on-snow workshops to create an experiential sense of glacier retreat that parallels the scientific data of ice disappearance.

Reconnection is also temporal and technological. Palomino and Katz used 3D scanning and AI to return fish-skin parkas, long held in metropolitan museums, to Arctic hands and eyes, advancing Indigenous intellectual sovereignty in digital space [6]. By contrast, Jokela worked in and with the Ounas River, to create art over decades and act as a lever for Arctic sustainability transformation. Through these efforts of reconnection, we can learn to stay with "trouble" as Haraway called it, through acts of stirring up or creating turbidity, in order to respond to environmental degradation of the environment through "response-ability" [7].

In weaving these threads, the collection contributes to a growing scholarship that sees aesthetics not as peripheral but as foundational infrastructure for environmental understanding and action [8]. The dialogue between Indigenous and non-Indigenous voices in this collection seeks to honor the specificity of place, knowledge, and history, and their rich diversity. As ice recedes and ecosystems shift, the Arctic will continue to demand new forms of witness and response. The chapters gathered here show that when art and scholarship meet—through collaboration, play, and sustained making—witness becomes action and documentation becomes dialogue.

These chapters ask us to remain attentive, to listen, to look closely, to engage deeply with the more-than-human world and its ongoing changes. Through practices of witness and observation, through art and science, we shape the ways in which environmental change is understood, remembered, and acted upon. As the Arctic warms at an accelerating pace, such acts of attention will be ever more vital. Ultimately, this collection affirms that we must collectively document and respond to the accelerating environmental transformations of our time, as we move in reciprocal dialogue with the living landscapes that we are of.

References

1. Abram, D. (1996). *The spell of the sensuous: Perception and language in a more-than-human world*. New York: Vintage.
2. Rose, D. B. (2013). Val Plumwood's philosophical animism: Attentive interactions in the sentient world. *Environmental Humanities*, 3, 93–109.
3. Ingold, T. (2011). *Being alive: Essays on movement, knowledge and description*. London: Routledge.
4. Kester, G. (2004). *Conversation pieces: Community and communication in modern art*. Berkeley: University of California Press.
5. Bartlett, C., Marshall, M., & Marshall, A. (2012). Two-eyed seeing and other lessons learned within a co-learning journey of bringing together Indigenous and mainstream knowledges. *Journal of Environmental Studies and Sciences*, 2(4), 331–340.
6. Smith, L. T. (2012). *Decolonizing methodologies: Research and indigenous peoples* (2nd ed.). London: Zed Books.
7. Haraway, D. (2016). *Staying with the trouble: Making Kin in the Chthulucene*. Durham, NC: Duke University Press.
8. Gabrys, J. (2016). *Program earth: Environmental sensing technology and the making of a computational planet*. Minneapolis: University of Minnesota Press.

Index

Pages in *italics* refer to figures.

Made in the USA
Las Vegas, NV
19 September 2026

293946fa-9004-4800-a458-39fc03f381afR02